Analog
Instrumentation
Fundamentals

Vincent F. Leonard, Jr., is currently Associate Professor of Electrical Engineering at Jamestown Community College, a State University of New York campus. Prior to coming to Jamestown he served as Department Chairman of the Aircraft Electronic Technology Program at the Academy of Aeronautics (LaGuardia Airport, New York). Professor Leonard has served as a consultant to local industries, has cosponsored several microcomputer workshops in the Jamestown area, and is a member of ASEE.

Analog Instrumentation Fundamentals

by
Vincent F. Leonard, Jr.

Associate Professor of Electrical Engineering
Jamestown Community College

Howard W. Sams & Co., Inc.
4300 WEST 62ND ST. INDIANAPOLIS, INDIANA 46268 USA

International Standard Book Number: 0-672-21835-6
Library of Congress Catalog Card Number: 81-51555

Edited by: *Jack Davis*
Illustrated by: *Ralph Lund*

Printed in the United States of America.

Preface

This book is about basic analog instruments—how they work, how to design them, and how to use them. Such information continues to be important despite the invasion of instrumentation by digital electronics. The reasons why this is so include the following:

- For persons studying to become electronics technicians the analysis of basic analog instruments provides a sound review (with applications) of basic circuit theory.
- Analog instruments *still have* some *advantages* over digital instruments. Consider, for example, the measurement of voltage in an environment of strong electromagnetic fields (radio stations). Digital voltmeters tend to lose counts very easily in such an environment, but analog voltmeters are not affected nearly so much. Also, if you wish to observe trends (is it increasing or decreasing?) it is usually much easier to observe the motion of a pointer than to watch a series of flashing numbers.
- Industry still uses large numbers of analog instruments.
- A foundation of basic analog principles is highly desirable prior to studying digital instrumentation. A knowledge of attenuators, filters, op amps, etc., is necessary for proper "signal conditioning" of signals that are to be processed digitally.

These, and certain more personal reasons, are why I wrote this book.

To understand the text only a knowledge of basic electronics and elementary algebra is required. Some of this background material has been included in the appendixes. The term "current" is used to mean electron flow, *not* "conventional current."

I *do hope* that you enjoy this book. If you occasionally encounter a little humor in the text, rest assured it was intended.

VINCENT F. LEONARD, JR.

Dedication

During the course of writing this book I reflected upon to what or to whom it should be dedicated. This is no small matter, and it consumed a considerable amount of my time. Thus I dedicate this scholarly work to the following:

1. The politicians and economists whose prudent fiscal policies made the creation of this work an absolute necessity.
2. *Most of all to:*
   a. My wife, Jeanne.
   b. My children, Vincent, Cynthia, Stephen, Darlene, and Tommy.
   c. My father, Vincent F. Leonard.

Special thanks are due the following:

1. Professor Andrew Staugaard for encouraging me to write, in a field that he wasn't.
2. Andy's lovely wife, Janet, who typed the text.
3. Jon and Chris Titus, Dave Larsen, and all the Blacksburg group.
4. Dr. Daniel Courtine, whose complete lack of self respect enabled him to agree to be used as an example of *what not to do* throughout the text.

Contents

APPENDIX E

APPENDIX F

CHAPTER 1

Fundamental Measurement Concepts

1-1 INTRODUCTION

When you are measuring a length, distance, height, or weight or other quantity, it is very important that you report your results based on a standard unit of measure. For this reason the *metric* and *British* measurement systems will be discussed in this chapter, along with the appropriate conversion factors. When discussing measurement results it is also important that you can discuss the accuracy and resolution of your measurement instruments, since this will affect the accuracy of your results. A number of other terms in measurement are important, including precision, error, range, and standards. A short history of the metric system of measurement will also be discussed, including the national and international organizations that are responsible for the maintenance of old standards and the creation of new standards.

1-2 OBJECTIVES

At the end of this chapter you will be able to do the following:

- Define the following terms:

measurement	sensitivity
instrument	standard
accuracy	analog signal

error   digital signal
precision   resolution
range

- Distinguish between international, primary, secondary, and working standards.
- Specify the unit name and symbol for fundamental international standard (SI) quantities.
- Specify the symbol and value of commonly used prefixes.
- Convert from one measurement system to another.
- Explain why an instrument can never be 100 percent accurate.

1-3 THE IMPORTANCE OF MEASUREMENT

In its most basic form the process of *measurement* involves comparing an unknown with a standard, in order to determine how large the unknown is, compared to the standard. More complex measurements often employ counting techniques and or statistical methods. For practical reasons the process of measurement requires a commonly agreed on system of units and standards. Thus the lengths of objects can be measured in inches, feet, or meters, and volume by quarts, liters, or bushels. Trade and commerce depend on a standardized system of weights and measures. *Instruments* are the tools used to make these measurements. Depending on the specific measurement the appropriate instrument may be very simple or extremely complex. Several "working definitions" of terms frequently encountered when measurement is discussed are given below:

accuracy—The degree of agreement (closeness) between the measured value and the actual value of the unknown.

error—Any deviation in the measured value from the actual value of the unknown.

precision—A measure of how *repeated* measurements (when the same units of measure are used) agree with each other. Precision should *not* be confused with accuracy.

range—The region bounded by the upper and lower limits of the instrument. For example, an ammeter might have a range specified as 0–100 mA.

resolution—The smallest change in measured value to which the instrument responds.

sensitivity—The smallest input that can produce a specified output.

standard—What the unknown is compared to. A standard is the *physical representation* of a unit of measurement.

Sir William Thomson (1824–1907), better known as Lord Kelvin, was interested in the practical application of scientific knowledge. Lord Kelvin made significant contributions to such diverse fields as thermodynamics, navigation, atomic theory, and submarine cable telegraphy. In conjunction with the last interest, Lord Kelvin invented the mirror galvonometer and siphon recorder. These instruments detected the presence of very small electrical currents—thus making practical the reception of signals transmitted over submarine cables. Lord Kelvin understood the importance of measurement as the following quotation illustrates:[1]

"I often say that when you can measure what you are speaking about and express it in numbers, you know something about it; but when you cannot express it in numbers, your knowledge is of a meagre and unsatisfactory kind; it may be the beginning of knowledge, but you have scarcely, in your thoughts, advanced to the stage of science, whatever the matter may be."

Lord Kelvin's words are as relevant today as they were in his time. One of our goals in this book is to go beyond "the beginning of knowledge." This journey will enable us to understand how our instruments work, their limitations, and the role they play in our lives. The trip may not always be easy but (it is hoped) it will never be more difficult than it has to. For the serious student or hobbyist it is a journey well worth taking.

1-4 UNITS AND STANDARDS

The development of the metric system was largely a result of the French Revolution. The revolutionaries, realizing the desirability of a system of standard weights and measures, posed the practical problem of developing such a system to the scientists. Thus in 1790 the Paris Academy of Sciences established a committee composed of Laplace, Lagrange, Lavoisier, and other distinguished scientists to propose the desired system.[2] These eminent scientists decided to base the system on *permanent natural standards* so that it would be applicable for all time. Thus the standard of length (meter) was initially defined as one ten-millionth of the distance from the earth's equator to either pole, along any meridian.[3] A standard meter bar (a unit of length) and standard kilogram (a unit of mass) made of platinum were constructed and adopted as the basis of all measurements in France in 1799. These initial standards remained in France where they were carefully stored. Due to measurement errors, these standards did not have the exact values intended. For practical reasons, however, standards constructed later were based on the initial French standards. In recent times new standards have been proposed (and in some instances adopted) based on the original concept of permanent natural standards. This has the advantage of greater permanence for the standards and, possibly, greater precision.[4]

The popularity of the French metric system created a demand for copies of the French standards and an international organization to ensure their uniformity. Thus in 1875 a treaty (Convention of the Meter) was signed, creating a permanent organization—the structure of which is summarized below:[5]

1. *General Conference on Weights and Measures*
This body meets at six-year intervals and includes delegates from all member countries. The General Conference is the ultimate authority.

2. *International Committee on Weights and Measures*
This committee consists of 18 members (each from a different country). The committee considers problems referred to it by the general conference and supervises the operations of the bureau discussed below.

3. *International Bureau of Weights and Measures*
The Bureau is located at Sèvres, France, and is the workhorse of the organization. Its responsibilities include maintaining the international standards, checking the national standards of member countries, and engaging in research to improve standards and measurement techniques.

In addition to the International Bureau of Weights and Measures numerous *national laboratories* have been established by various countries. In the United States the *National Bureau of Standards* (NBS), located in Washington, was created in 1901. Thus numerous standards exist today, forming the *hierarchy of standards* outlined below:

1. *International standards* maintained by the International Bureau of Weights and Measures.

2. *National* (primary) *standards* maintained by the various national laboratories.

3. *Secondary standards,* which are the fundamental standards maintained and *used by industry*. Frequently these standards are sent to a national laboratory for calibration with the National Standards.

4. *Working standards,* which are used by industry for the *routine calibration* of instruments, parts, etc. From time to time the working standards are compared to the secondary standards.

Although the United States signed the Convention of the Meter in 1875, it chose to utilize the metric system *only* in international transactions. Thus the familiar and rather archaic British system (feet and pounds) is still used domestically. In the not too distant future the metric system should replace the British system in the United States.

In 1960 the Eleventh General Conference of Weights and Measures adopted the SI (*Système International d'Unités*) system. This system is gaining widespread acceptance in industry and thus will be emphasized in this text. Table 1-1 lists the *fundamental* SI quan-

Table 1-1. Fundamental SI Quantities and Units

Quantity	Unit Name	Unit Symbol
length	meter	m
mass	kilogram	kg
time	second	s
current	ampere	A
temperature	degree kelvin	°K
luminous intensity	candela	cd

tities and units. Similarly Table 1-2 lists the commonly used *prefixes*. Note in Table 1-2 that commas are *not* used to separate groups of three digits—a space is used instead. Obviously many more units exist than those listed in Table 1-1. Such units are called *derived units* since they are expressed (via a defining equation) in terms of the six fundamental units listed in Table 1-1. Examples of derived SI units are provided in Table 1-3. The Eleventh General Confer-

Table 1-2. Commonly Used Prefixes

Prefix	Symbol	Factor	Scientific Notation Equivalent
tera	T	1 000 000 000 000	10^{12}
giga	G	1 000 000 000	10^{9}
mega	M	1 000 000	10^{6}
kilo	k	1 000	10^{3}
milli	m	0.001	10^{-3}
micro	μ	0.000 001	10^{-6}
nano	n	0.000 000 001	10^{-9}
pico	p	0.000 000 000 001	10^{-12}

ence designated the units for plane-angle measure (radian) and solid-angle measure (steradian) as "supplementary." Note in Table 1-3 that the definition of the lumen is given in terms of the supplementary solid-angle unit.

Since the British system is still employed (for domestic use) in certain remote sections of the world, appropriate conversion factors are provided in Appendix A.

Table 1-3. Derived SI Units

Quantity	Unit Name	Unit Symbol	Definition
force	newton	N	$kg \cdot m/s^2$
work, energy	joule	J	$N \cdot m$
power	watt	W	J/s
pressure	pascal	Pa	N/m^2
charge	coulomb	C	$A \cdot s$
voltage	volt	V	W/A
resistance	ohm	Ω	V/A
inductance	henry	H	$V \cdot s/A$
capacitance	farad	F	$A \cdot s/V$
frequency	hertz	Hz	s^{-1}
magnetic flux	weber	Wb	$V \cdot s$
magnetic flux density	tesla	T	Wb/m^2
luminous flux	lumen	lm	$cd \cdot sr$
illumination	lux	lx	lm/m^2

1-5 CONVERSION OF UNITS

It is desirable to develop the ability to efficiently manipulate (via scientific notation) the prefixes given in Table 1-2. In addition, it is sometimes necessary to convert from one system of measure to another. A brief review of scientific notation is provided in Appendix B for those readers who may need some "brushing up" on this notation. The following discussion and example problems illustrate the method used to convert from one system of measure to another.

We wish to convert some unit A (employed in one system) to another unit B (employed in another system). To begin, note that a certain number of A units (N_1) equals a certain number of B units (N_2). Thus

$$N_1 \times A = N_2 \times B$$

Therefore

$$\frac{N_1 A}{N_2 B} = \frac{N_2 B}{N_1 A} = 1$$

Since multiplication by 1 introduces no net change, we have a simple means to convert units when the need arises. Specifically:

To convert from A units to B units we multiply $N_1 A$ by $N_2 B/N_1 A$.
To convert from B units to A units we multiply $N_2 B$ by $N_1 A/N_2 B$.

EXAMPLE 1-1

One inch equals (approximately) 2.54 cm. Convert the following to centimeters:
(a) 6.5 in
(b) 1 ft, 3 in (1 ft = 12 in).

(*a*) $1 \text{ in} = 2.54 \text{ cm}$. Note that $1 = N_1$, $\text{in} = A$, $2.54 = N_2$ and $\text{cm} = B$. Therefore

$$\frac{1 \text{ in}}{2.54 \text{ cm}} = \frac{2.54 \text{ cm}}{1 \text{ in}} = 1$$

$$6.5 \text{ in} \times \frac{2.54 \text{ cm}}{1 \text{ in}} = 16.51 \text{ cm}$$

(*b*) $1 \text{ ft, } 3 \text{ in} = 15 \text{ in}$. Therefore

$$15 \text{ in} \times \frac{2.54 \text{ cm}}{1 \text{ in}} = 38.1 \text{ cm}$$

EXAMPLE 1-2

An emporium of libation (beer hall) is 36 ft long by 24 ft wide. Determine the floor area in square meters.

$$A = 36(24) = 864 \text{ ft}^2$$
$$1 \text{ ft}^2 = 929 \text{ cm}^2 \text{ (Appendix A)}$$
$$\frac{1 \text{ ft}^2}{929 \text{ cm}^2} = \frac{929 \text{ cm}^2}{1 \text{ ft}^2} = 1$$
$$864 \text{ ft}^2 \times \frac{929 \text{ cm}^2}{1 \text{ ft}^2} = 802\,656 \text{ cm}^2$$

To convert square centimeters (cm^2) to square meters (m^2), note that $1 \text{ m} = 100 \text{ cm}$. Therefore

$$1 \text{ m}^2 = (100 \text{ cm})^2 = 1 \times 10^4 \text{ cm}^2$$
$$\frac{1 \text{ m}^2}{1 \times 10^4 \text{ cm}^2} = \frac{1 \times 10^4 \text{ cm}^2}{1 \text{ m}^2} = 1$$
$$802\,656 \text{ cm}^2 \times \frac{1 \text{ m}^2}{1 \times 10^4 \text{ cm}^2} = 802\,656 \times 10^{-4} \text{ m}^2$$
$$= 80.266 \text{ m}^2$$

Since Appendix A did not provide a conversion factor to convert square feet (ft^2) to square meters (m^2) we *derived* the needed information in the course of solving the problem. This is often the case in the "real world" since tables of conversion factors are rarely complete.

EXAMPLE 1-3

A rather outstanding electronics student is described by the following dimensions: 96.52 cm, 60.96 cm, 91.44 cm. In order to practice converting from one system of units to another convert the given dimensions to hands, where 1 hand (hd) = 4 inches.

Considering the first dimension,

$$\frac{1 \text{ in}}{2.54 \text{ cm}} = \frac{2.54 \text{ cm}}{1 \text{ in}} = 1$$

$$96.52 \text{ cm} \times \frac{1 \text{ in}}{2.54 \text{ cm}} = 38 \text{ in}$$

$$38 \text{ in} \times \frac{1 \text{ hd}}{4 \text{ in}} = 9.5 \text{ hds}$$

Often it is quicker to perform multiple conversions in a "chain fashion" as illustrated with the remaining dimensions:

$$60.96 \text{ cm} \times \frac{1 \text{ in}}{2.54 \text{ cm}} \times \frac{1 \text{ hd}}{4 \text{ in}} = 6 \text{ hds}$$

$$91.44 \text{ cm} \times \frac{1 \text{ in}}{2.54 \text{ cm}} \times \frac{1 \text{ hd}}{4 \text{ in}} = 9 \text{ hds}$$

Thus

$$96.52 \text{ cm} = 38 \text{ in} = 9.5 \text{ hds}$$
$$60.96 \text{ cm} = 24 \text{ in} = 6 \text{ hds}$$
$$91.44 \text{ cm} = 36 \text{ in} = 9 \text{ hds}$$

Clearly the dimensions are *outstanding* regardless of which system of units is employed!

1-6 THE CERTAINTY OF UNCERTAINTY

No matter how sophisticated, well designed, or carefully constructed an instrument is, it can *never* produce a measured value of the unknown that is exact. A little thought will reveal numerous sources of uncertainty associated with the measurement process. A few of the more obvious sources of uncertainty include the inability of people to read scales exactly, inherent instrument errors due to the uncertainties in their comparison with standards, instrument components that change with time or temperature, and so on. Even if the instrument used for a particular measurement were perfect, there would still be a degree of uncertainty associated with the value of the unknown! This uncertainty exists because *the act of measurement changes the quantity being measured* in such a manner that it is not possible to determine the value of the unknown "exactly." Obviously, then, you wish to limit the changes introduced by your instruments to very small values in order to obtain accurate values for the unknown. This is relatively easy to do in the macroscopic world. Thus, while the uncertainties in your measurements can never *be* zero they can be made to *approximate* zero. In the ultra-microscopic atomic world it is apparently *not possible* to devise instruments that introduce only very small changes. To understand why this is so we will briefly discuss the Heisenberg uncertainty principle and make some "ball park" calculations using sensible approximations.

In 1927 Werner Heisenberg published a paper in the German periodical *Physikalische Zeitschrift* in which he stated his uncertainty principle.[6] This principle states that it is impossible to obtain simultaneously exact values for both the position and momentum of a particle. Specifically,

$$\Delta x \, \Delta p \geqq \frac{h}{2\pi} \qquad (1\text{-}1)$$

where
$\Delta x =$ uncertainty in the measurement of position, which is in centimeter units,
$\Delta p =$ uncertainty in the measurement of momentum, which is simply the product of mass (units of grams) and velocity (units of centimeters per second),
$h =$ Planck's constant, $6.625\,6 \times 10^{-27}$ erg-second.

The uncertainties in Equation 1-1 are *not* due to imperfections of the measuring instruments! Matter and energy have definite limits to their "smallness." To determine information about the position of a particle, light or some other form of electromagnetic energy is employed. Since the smallest unit of light energy is a photon, this represents the smallest amount of light that can be used to measure the position of a particle. Now if the particle being measured is itself very small then the interaction with even a single photon will introduce drastic uncertainties. Thus, if you attempt to accurately measure position, the act of measurement changes momentum, and vice versa. In order to appreciate the magnitude of uncertainty introduced by such measurements Equation 1-1 is used:[7]

$$\Delta x \, \Delta p = \frac{h}{2\pi}$$

Substituting mass times velocity (mv) for p gives

$$\Delta x \, \Delta mv = \frac{h}{2\pi}$$

If the velocity of the particle (v) is small compared with the velocity of light ($2.997\,9 \times 10^{10}$ cm/s) then little error is introduced by assuming that m is constant. Thus

$$\Delta x \, m\Delta v = \frac{h}{2\pi}$$

Letting the numerical value of Δx and Δv be equal, which makes the uncertainties in position and velocity equal,

$$\Delta x \, m\Delta x = \frac{h}{2\pi}$$

$$m \, (\Delta x)^2 = \frac{h}{2\pi}$$

Finally, solving for the uncertainty in position, Δx,

$$(\Delta x)^2 = \frac{h}{2\pi m} = \frac{6.625 \, 6 \times 10^{-27}}{2\pi m}$$

$$= \frac{1.054 \times 10^{-27}}{m} \cong \frac{1 \times 10^{-27}}{m}$$

$$\Delta x \cong \sqrt{\frac{1 \times 10^{-27}}{m}} \cong \frac{3.2 \times 10^{-14}}{\sqrt{m}} \qquad (1\text{-}2)$$

where $\sqrt{m}$ is the square root of the object's mass.

Equation 1-2 enables you to *estimate* the amount of uncertainty (in position and velocity) which results when you attempt to measure the position of an object. All you need do is substitute the mass in grams of the object into Equation 1-2 and calculate the resulting uncertainty. Dr. Isaac Asimov suggests adjusting the unit to the object, by taking the diameter of the object as the measure of position. He has done this for a number of objects, and the results of his calculations are summarized in Table 1-4.

Table 1-4. Asimov's Uncertainty Calculations

Object	Approximate Diameter (cm)	Approximate Mass (g)	Uncertainty in Position (cm)	Uncertainty in Position (Diameters)
ameba	1.6×10^{-2}	4×10^{-6}	1.6×10^{-11}	1×10^{-9}
bacterium	1×10^{-4}	1×10^{-12}	3.2×10^{-8}	3.2×10^{-4}
gene	3.4×10^{-6}	4×10^{-17}	5×10^{-6}	1.5
uranium atom	1×10^{-8}	4×10^{-22}	1.6×10^{-3}	1.6×10^5
proton	1×10^{-13}	1.6×10^{-24}	2.5×10^{-2}	2.5×10^{11}
electron	1×10^{-13}	9.1×10^{-28}	1.1	1.1×10^{13}

EXAMPLE 1-4

Verify the uncertainties in position given in Table 1-4 for (a) an ameba and (b) an electron. Discuss the results.

(a)

$$\Delta x = \frac{3.2 \times 10^{-14}}{\sqrt{m}} = \frac{3.2 \times 10^{-14}}{\sqrt{4 \times 10^{-6}}}$$

$$= \frac{3.2 \times 10^{-14}}{2 \times 10^{-3}} = 1.6 \times 10^{-11} \text{ cm}$$

In diameters,

$$\Delta x = 1.6 \times 10^{-11} \text{ cm} \times \frac{1 \text{ diameter}}{1.6 \times 10^{-2} \text{ cm}} = 1 \times 10^{-9}$$

(b)

$$\Delta x = \frac{3.2 \times 10^{-14}}{\sqrt{m}} = \frac{3.2 \times 10^{-14}}{\sqrt{9.1 \times 10^{-28}}}$$
$$= \frac{3.2 \times 10^{-14}}{3.017 \times 10^{-14}} = 1.1 \text{ cm}$$

In diameters,

$$\Delta x = 1.1 \text{ cm} \times \frac{1 \text{ diameter}}{1 \times 10^{-13} \text{ cm}} = 1.1 \times 10^{13}$$

The uncertainty in measuring the position of an ameba is approximately 1.6×10^{-11} cm (0.000 000 000 016 cm) or 1×10^{-9} diameters (0.000 000 001 diameters). Obviously the amount of uncertainty is so small that you can feel quite confident in being able to measure accurately the position of an object the size of an ameba.

The uncertainty in measuring the position of an electron is 1.1 cm or 1.1×10^{13} diameters (11 000 000 000 000 diameters). This is a *huge uncertainty!* The uncertainty principle tells you that even with a perfect instrument it is virtually impossible to measure accurately the position of an object the size of an electron. Fortunately, the kinds of instruments that will be discussed in this text will be used to measure quantities in the macroscopic world. Thus we need not further consider the types of uncertainties illustrated in Example 1-4. The concept that *will* be retained from Example 1-4, however, is this: it is *impossible* to devise an instrument that is 100-percent accurate.

1-7 ANALOG AND DIGITAL SIGNALS

We live in an analog world where the majority of signals detected by our senses are analog in nature. An *analog signal* is one that can assume any value within the range of possible values in a given system. Our eyes can detect light even though it may not be very intense (moonlight or starlight). Our bodies can also sense small changes in temperature and pressure, and most of us can tell what we are having for dinner before even entering the kitchen.

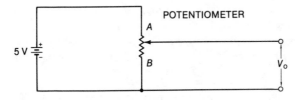

Fig. 1-1. A simple analog voltage generator.

Consider, for example, the power supply shown in Fig. 1-1. By varying the position of the potentiometer wiper, any output voltage between 0 V and 5 V can be obtained. If the wiper is moved between A and B in uniform manner, the analog voltage shown in Fig. 1-2 would be generated. An analog system can be almost anything—an electronic circuit, living organism, chemical process, etc. Analog signals are frequently described as *continuously changing*. This does

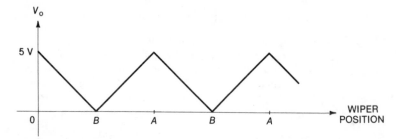

Fig. 1-2. Moving the potentiometer wiper (Fig. 1-1) in a uniform manner.

not mean that at a particular instant an analog signal must be changing, but rather that when the analog signal does change it normally makes a smooth transition from one value to another. Thus, during the time interval an analog signal changes from one value to another; it will, at different instants in time, assume *all values* between the initial and final values.

The instruments you will learn about in this book are analog instruments. This means the instrument employs analog circuits and normally displays the result in an analog manner. The word "analog" is derived from "analogous," which means "similar to." Thus if we say A is the analog of B we mean that A changes in a manner that is similar (analogous) to the way B changes. The following examples of common analog relationships and associated instruments should clarify this concept:

- *Thermometer*—The height of a column of mercury is analogous to temperature.
- *Automobile speedometer*—The speed of an automobile (indicated by the amount of deflection of a pointer) is analogous to the rate of rotation of the drive shaft.
- *Light meter*—Light intensity (indicated by the amount of deflection of a pointer) is analogous to the amount of current through a phototransistor.

Digital signals are different from analog signals. A *digital signal* is one that can assume only one of two possible values. For example,

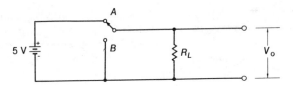

Fig. 1-3. A simple digital voltage generator.

consider the power supply shown in Fig. 1-3. The output voltage is either 5 V (position A) or 0 V (position B). Output voltages between 0 V and 5 V are not possible. If the switch is moved back and forth between A and B, the digital voltage shown in Fig. 1-4 is generated. When digital signals vary they do so in *discrete* steps. Thus when a digital signal changes from one value to another the change is *discontinuous* (abrupt). Digital instruments utilize circuits that operate with only two voltage levels. Consequently, in order for a digital instrument to measure an analog signal it must first convert the analog signal to a digital signal. This conversion is accomplished by an analog to digital converter (A/D converter). Advances in integrated-circuit (IC) technology have dramatically decreased the cost

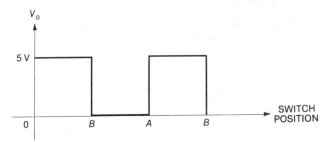

Fig. 1-4. Generating either 0 V or 5 V from a switch and battery (Fig. 1-3).

of digital instruments. Because of this, many analog instruments are gradually being replaced by digital instruments. While digital instruments have many advantages over analog instruments, they also have some disadvantages. Thus digital instruments are not likely to completely replace analog instruments in the near future.

1-8 REVIEW OF OBJECTIVES

In this chapter you learned the definitions of several terms commonly used with instruments. The development of the metric system, and an introduction to the SI system of units was provided. In addition, you learned about international and national organizations which define, maintain, and strive to improve standards, techniques of measurement, and instruments. The use of prefixes and a technique for converting from one system of measure to another was discussed. Heisenberg's uncertainty principle was introduced, and the difference between analog and digital signals was discussed.

1-9 QUESTIONS

1. Explain the difference between precision and accuracy.
2. What is meant by a hierarchy of standards?
3. What functions does the NBS serve?
4. What advantages to the United States would result from converting to the metric system? Disadvantages?
5. What is the difference between a working standard and a secondary standard?
6. What is a derived unit of measure?
7. Does the Heisenberg uncertainty principle significantly limit the accuracy of most measurements? Why?
8. What is the name of the periodical in which Heisenberg published his famous paper? Why do you think this is *not* a household word in Jamestown, New York?
9. A colleague of the author (Dr. Daniel Courtine) recently proposed the adoption of a new unit of length (the courtine) for neurosurgeons. The courtine is defined as the distance between two wide lines on a clay tablet kept on Dr. Courtine's window sill. Why is it unlikely the proposed unit will gain widespread acceptance?

1-10. PROBLEMS

1. Convert the following:
   23 μV to millivolts   0.000 085 MΩ to kilohms
   1 300 mV to microvolts   236 000 KΩ to megohms
   15 pF to microfarads   1 212 μA to milliamperes
2. Determine the area of a wall 9 ft $\times$ 20 ft in square meters.
3. The student in Example 1-3 has a mass of 110 lbm. Convert this mass to kilograms.
4. Assuming equal uncertainties in position and momentum, determine the uncertainty of position in diameters for a particle 1×10^{-10} m in diameter, possessing a mass of 3.5×10^{-20} g.
5. What is the approximate diameter of an ameba in inches?
6. Which SI unit does the product of resistance and capacitance yield? Inductance divided by resistance?
7. Perform the following operations. Express answers using standard prefixes, and units. If necessary, refer to Appendix A.
   25 mA $\times$ (40 kΩ)   1 000 μV/25 mΩ
   25 μA $\times$ (4 MΩ)   500 mA $\times$ (5 Ω)
   125 mV/4 MΩ 2 000 μV/20 pA
8. A motor delivers 5.94×10 ft·lbf of power to a load. What is the horsepower (hp) rating of the motor? The wattage?
9. What is the mass, in tons, of an object whose mass is 5.28×10^6 kg?

1-11 REFERENCES

1. Weber, Robert L.; Manning, Kenneth V.; and White, Marsh W. 1965. *College Physics.* New York: McGraw Hill, p. 9.
2. Mason, Stephen F. 1977. *A History of the Sciences.* New York: Collier Books, p. 435.
3. Meyer, Jerome S. 1962. *The ABC of Physics.* New York: Pyramid Publications, p. 13.
4. Crittenden, E. C. 1969. "Metric System," *Encyclopedia Americana,* XVIII, p. 729.
5. Ibid., p. 730.
6. Gamow, George. 1966. *Thirty Years That Shook Physics.* New York: Anchor Books, p. 98.
7. Asimov, Isaac. 1972. *From Earth to Heaven.* New York: Discus/Avon Books, pp. 147–52. The chapter entitled "The Certainty of Uncertainty" seemed so appropriate that I chose it for the title of Section 1-4. Readers interested in learning more about uncertainty are encouraged to acquire Dr. Asimov's book. Readers who are uncertain about the desirability of exploring the uncertainty principle in more detail are certainly encouraged to also acquire a copy of this excellent book.

CHAPTER 2

Analog Meter Movements

2-1 INTRODUCTION

In this chapter you will learn how analog meter movements work. The principles of operation, construction details, and specifications for a number of analog meter movements are provided.

Since the d'Arsonval and taut-band movements are by far the most common movements, we will discuss these movements in detail. In addition, you will learn how to experimentally determine the input resistance of a meter movement.

2-2 OBJECTIVES

At the end of this chapter you will be able to do the following:

- List some properties of an ideal meter movement.
- Explain how a real meter movement differs from an ideal meter movement.
- Interpret the following specifications: input resistance, accuracy, sensitivity, and damping characteristics.
- Discuss in some detail the construction of d'Arsonval and taut-band meter movements.
- Explain the principle of operation of the iron-vane and electro-dynamometer meter movements.
- Determine the internal resistance of a meter movement.

2-3 METER MOVEMENTS

The "heart" of an analog instrument is the meter movement employed in that instrument. A *meter movement* is an elementary *current*-measuring device (galvanometer). The principle of operation of most meter movements is based on the principle that *a force is exerted on a current-carrying conductor placed in a magnetic field.* The amount of current through a meter movement is indicated by the amount of deflection of a pointer, along a calibrated scale. If it were possible to construct an ideal meter movement, some of the properties it would have to include are the following:

1. *Zero resistance.*
2. The amount of *deflection* of the pointer would be *directly proportional to* the amount of *current* through the meter movement. This means that if the amount of current through the meter movement changes by a certain amount (X percent), then the amount of deflection indicated by the pointer would also change by *exactly* the same amount (X percent).
3. *100-percent accuracy.* The measured value of current would equal the actual value of current.
4. *100-percent precision.* Repeated measurements would always be the same.
5. *Infinite range.* The meter movement could measure any value of current.
6. *Ideal sensitivity.* The smallest possible change in current would produce a readable change in pointer deflection. The smallest known charge is the amount of charge associated with an electron or proton ($\pm 1.602 \times 10^{-20}$ emu). Thus the ideal meter movement could detect a change of 1.602×10^{-20} emu per unit time!

Other properties such as durability, ruggedness, low cost, etc., could easily lengthen the list of ideal properties given above. Obviously a meter movement with the characteristics listed above *cannot* be constructed. In fact, for certain measurements some of the ideal properties would be undesirable! For example, if a meter movement were too sensitive, the pointer would never come to rest, due to random movements of very small charges within the conductor. How do real meter movements differ from ideal meter movements? Real meter movements can only *approximate* the properties listed for ideal meter movements. How closely a real meter movement approximates an ideal meter movement is a measure of the overall quality (and often cost) of the meter movement. Similarly, an instrument employing a meter movement can be no better than the meter movement it is designed around!

2-4 SPECIFICATIONS

Real meter movements have a finite amount of resistance (R_m). The smaller the resistance of a meter movement, the better. When a meter movement is used to measure current it is connected in *series* with the circuit through which the current is flowing. Therefore, to prevent the meter movement from disturbing the circuit its internal resistance (R_m) should be small compared with the resistance of the circuit in which it is being used. More will be said about this concept when we discuss "loading effects." For now the idea is simply this: *it is desirable to select a meter movement that has a low internal resistance.*

Real meter movements do *not* produce deflections that are *exactly* proportional to current. The scale of a meter movement is *linear*, meaning it is marked with equally spaced (uniform) divisions. The scales for meter movements are usually mass produced. Thus different individual meter movements of the same type and manufacturer have essentially the same scale. Since there are slight variations in the characteristics from one meter movement to the next, different meter movements will *not* indicate *exactly* the same reading for the same current. Some meter movements will read slightly high, others slightly low. Consequently, meter movements have a certain amount of *calibration error* because the individual units have not been separately calibrated. The reason that most meter movements are not separately calibrated is simply due to the fact that the cost of the meter movement would increase significantly. Clearly, a real meter movement cannot be 100-percent accurate. It is interesting to note that the *accuracy* of a meter movement, *as it is stated in its specifications, is expressed in terms of error.* Accuracy is normally stated as a *percentage of full-scale deflection.* It is important to understand the implications of this type of specification. Let us proceed eagerly to Example 2-1.

EXAMPLE 2-1

The accuracy of a 0- to 100-μA meter movement is stated as 3 percent of the full-scale value. Determine:
  (*a*) The amount of calibration error.
  (*b*) The range in actual current for the following measured currents: 100, 90, 80, 70, 60, 50, 40, 30, 20, and 10 μA.
  (*c*) Error as a percentage of the measured current.
  (*d*) Discuss the results.

(*a*) 3 percent of 100 μA $= \dfrac{3}{100}(100\ \mu\text{A})$

Thus the calibration error (CE) is 3 μA.

(*b*), (*c*) To illustrate the significance of the given specification we provide detailed calculations for measured currents of 100 μA and 30 μA. For any value of measured current

$$R = I_m \pm CE \qquad (2\text{-}1)$$

where

R = range in actual current,
I_m = measured current,
CE = calibration error.

For a measured current of 100 μA,

$$R = 100 \ \mu A \pm 3 \ \mu A$$
$$= 97 \ \mu A \text{ to } 103 \ \mu A$$

This means the meter movement is "within specs" if it reads anywhere between 97 and 103 μA *when* the actual current through it is 100 μA.

The error as a percentage of the measured current (E_m) is given by

$$E_m = \frac{CE}{I_m} \times 100\% \qquad (2\text{-}2)$$

For a measured current of 100 μA

$$E_m = \frac{3 \ \mu A}{100 \ \mu A} \times 100\% = 3\%$$

Now for a measured current of 30 μA

$$R = 30 \ \mu A \pm 3 \ \mu A$$
$$= 27 \ \mu A \text{ to } 33 \ \mu A$$

Thus the meter movement is "within specs" if it reads anywhere between 27 and 33 μA *when* the actual current through it is 30 μA.

For this measurement the error as a percentage of the measured current of 30 μA is

$$E_m = \frac{3 \ \mu A}{30 \ \mu A} \times 100\% = 10\%$$

The results of similar calculations for the other measured currents are summarized in Table 2-1, where I_m is the measured current, CE the calibration error, R the range in actual current for the meter movement to be "within specs," and E_m the error as a percentage of measured current.

(d) Manufacturers state the accuracy of a meter movement as a percentage of full-scale deflection. Fig. 2-1 illustrates error as a percentage of measured current, for a 0- to 100-μA meter movement, which has an accuracy of 3 percent of the full-scale value. Notice in Fig. 2-1 (or Table 2-1) that although the calibration error (3 μA) is 3 percent of the full-scale value of 100 μA it is a whopping 30 percent of a measured

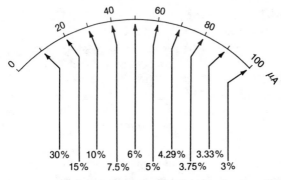

Fig. 2-1. Error as a percentage of measured value for the 0- to 100-μA meter movement in Example 2-1.

current of 10 μA! Thus *the effective error is not a fixed percentage of the measured value.* Clearly the effects of calibration error are most pronounced for "down-scale" readings. Therefore, when you use a meter movement to measure current, you should select one that produces "up-scale" readings for the range of currents you expect to measure. This practice, if followed, will minimize the adverse effects of calibration error.

Table 2-1. Summary of Calculations for Example 2-1

I_m (μA)	CE (μA)	R (μA)	E_m (%)
100	3	97–103	3
90	3	87–93	3.33
80	3	77–83	3.75
70	3	67–73	4.29
60	3	57–63	5
50	3	47–53	6
40	3	37–43	7.5
30	3	27–33	10
20	3	17–23	15
10	3	7–13	30

In Chapter 1 we defined *sensitivity* as the smallest input that can produce a specified output. Depending on the "input" and "specified output" given in the manufacturers' specifications, the following sensitivities can be defined:[1]

current sensitivity—The current in microamperes required to produce a deflection of one scale division.

voltage sensitivity—The voltage in microvolts that provides a deflection of one scale division when the meter movement is in series with a resistance equal to RCDRX. (For now assume

31

RCDRX is specified by the manufacturer. We will define RCDRX later.)

megohm sensitivity—The resistance in megohms that must be connected in series with the meter movement to produce a deflection of one scale division when the applied voltage is one volt.

ballistic sensitivity—The quantity of charge in microfarads per scale division required to momentarily deflect the instrument when it is used to measure the passage of a quantity of electricity.

Each of the definitions above satisfies the more general definition of sensitivity stated in Chapter 1. Depending on the intended use for the meter movement a particular definition would be preferred. For *our purposes* knowing the amount of *current* required to produce *full-scale* deflection (I_{FS}) will suffice, as an indication of a meter movement's sensitivity. Thus, a 50-μA meter movement will deflect full scale when the current through it is 50 μA. Similarly, a 1-mA meter movement deflects full scale when the current through it is 1 mA. Clearly, a 50-μA meter movement is *more sensitive* than a 1-mA meter movement. Commercially available meter movements typically provide full-scale deflections of 10 μA, 20 μA, 50 μA, 100 μA, 200 μA, and 1 mA. Generally speaking, the more sensitive a meter movement is, the better. The reason for preferring a more sensitive movement will become clear when we discuss voltmeters.

The operation of a basic meter movement is illustrated in Fig. 2-2. In Fig. 2-2A a small bar magnet is mounted on a pivot, be-

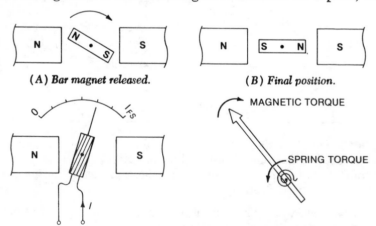

(A) *Bar magnet released.*

(B) *Final position.*

(C) *A pivoted coil "acts like" a bar magnet when an electric current flows through it.*

(D) *Spiral spring produces a counter torque which prevents the pointer from deflecting to the final position in (B).*

Fig. 2-2. Basic meter movement operation.

tween the poles of a larger permanent magnet. Assume you are holding the bar magnet in the position shown in Fig. 2-2A and then release it. Since like magnetic poles repel, a clockwise torque is produced that causes the bar magnet to rotate until it settles in the final position shown in Fig. 2-2B. In Fig 2-2C the bar magnet has been replaced with a pivoted coil, to which a pointer is attached. When current flows through the coil the ends of the coil become magnetized, producing north and south magnetic poles. Thus the current-carrying coil "acts like" the pivoted bar magnet in Figs. 2-2A and 2-2B. In order to prevent the coil and pointer from deflecting to the final position of Fig. 2-2B spiral springs are attached to the shaft that the coil and pointer are mounted on. This is illustrated in Fig. 2-2D. Thus the pointer will deflect to a point where the clockwise magnetic torque is exactly balanced by the counterclockwise spring torque. When there is no current through the coil the springs return the coil and pointer to the zero position. The amount of deflection is (approximately) proportional to the amount of current (I) through the coil. Thus the scale can be calibrated to read current directly.

If the current through the coil is suddenly removed, the pointer will of course swing towards the zero position. Because of the inertia of the coil assembly the pointer will move past the zero position and oscillate about it, for some time, before coming to rest. Similarly, if current is suddenly applied to the coil the pointer will oscillate briefly around the measured value. The *period* of the meter movement is the amount of time between consecutive passages of the pointer in the same direction through its zero position. The period depends on the stiffness of the spring and the mass of the coil and pointer. Increasing the stiffness reduces the period (which is desirable). If, however, the stiffness is increased, the sensitivity is reduced (which is undesirable) since a larger magnetic torque and hence more current is required to deflect the pointer. Decreasing the mass of the coil and pointer will also decrease the period. Thus it is desirable for a meter movement to have a small period so that the pointer will settle quickly on the measured value, rather than oscillating around it. The methods employed to minimize the period are called *damping* techniques. These techniques can be based on electrical principles, mechanical principles, or a combination of the two. When we discuss specific types of meter movements both electrical and mechanical damping will be employed. The damping characteristics of a meter movement can be classified as follows:

1. *Critically* damped. The pointer returns to the zero position in a minimum amount of time without swinging past zero ($R = $ RCDRX).

2. *Overdamped.* The pointer approaches zero sluggishly. $R <$ RCDRX (we will define RCDRX shortly).

3. *Underdamped.* The pointer swings past zero and tends to oscillate about it for a short time ($R >$ RCDRX).

The three cases of damping are illustrated in Fig. 2-3. Neglecting the resistance of the meter movement you will note that the battery

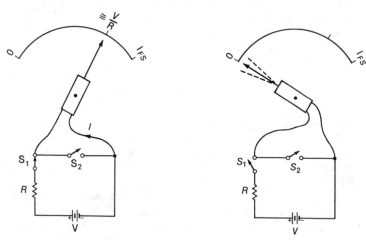

(A) When S_1 is opened the pointer swings past zero and then oscillates about it (underdamped).

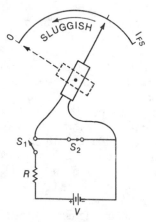

(B) At the instant S_1 is opened S_2 is closed, and the pointer approaches zero sluggishly (overdamped).

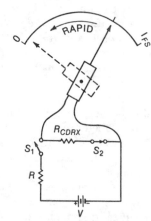

(C) At the instant S_1 is opened S_2 is closed and the pointer rapidly approaches zero without swinging past it (critically damped).

Fig. 2-3. Damping characteristics.

voltage V and resistance R establish a current through the meter movement equal to V/R. If the terminals of the meter movement are *open circuited* (Fig. 2-3A), the pointer will swing past zero and then oscillate about it. However, if the terminals of the meter movement are then *short circuited* (Fig. 2-3B), the pointer will sluggishly approach zero. As the coil and pointer swing towards zero they cut the magnetic lines of force produced by the permanent magnet. From basic electricity we know that whenever a conductor moves through a magnetic field a voltage is induced in the conductor.* Thus as the coil swings toward zero a voltage will be induced in it. If the coil is open circuited, the induced voltage cannot establish an induced current in the coil. Thus the motion of the coil is impeded only by the friction present in the system, producing the response illustrated in Fig. 2-2A. When the coil is short circuited, the induced voltage causes an induced current through the coil. Lenz's law tells us that the induced current will *oppose* the motion that induced it. Thus the magnetic field produced by the induced coil current opposes the magnetic field produced by the permanent magnet. The net effect of the interaction between the magnetic fields produced by the permanent magnet and current carrying coil is to produce a *braking action* which results in the coil and pointer moving sluggishly towards zero (Fig. 2-3B). The damping characteristics can be controlled by changing the value of the induced current. Thus for any meter movement a value of resistance exists which when connected across the coil produces the response illustrated in Fig. 2-3C (critically damped). This resistance is called the external critical damping resistance (RCDRX).[2] Critical damping represents the *desired damping characteristic* for a meter movement. In practice most meter movements are slightly underdamped.

2-5 THE D'ARSONVAL MOVEMENT

In 1882 the French scientist d'Arsonval Deprez developed a meter movement which proved to be the forerunner of the type of meter movements employed in modern analog instruments. The details of a typical d'Arsonval movement are illustrated in Fig. 2-4. You will note that the d'Arsonval movement is similar to the basic meter movement discussed previously. It consists of a coil that is free to rotate between the poles of a permanent horseshoe magnet (only the poles of the horseshoe magnet are illustrated in Fig. 2-4). Such an arrangement is referred to as a permanent-magnet moving-coil (pmmc) instrument. The major parts and functions of the d'Arsonval movement are discussed below.

---

* Assuming the conductor is *not* moving parallel to the magnetic lines of force.

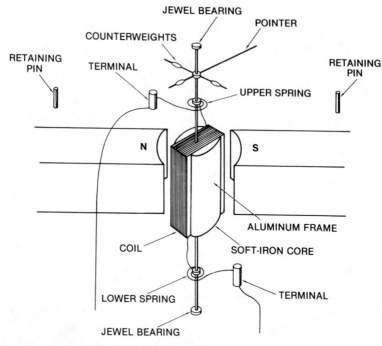

Fig. 2-4. The d'Arsonval meter movement.

Permanent Magnet

As illustrated in Fig. 2-4, the ends of the horseshoe magnet are fitted with *two curved pole pieces.* The purpose of the curved pole pieces is to concentrate the magnetic field produced by the permanent magnet in the space occupied by the moving coil. Typically the permanent magnet is made from an alloy of aluminum, nickel, and cobalt (Alnico). Alnico magnets provide a strong uniform magnetic field. In addition, Alnico magnets can be manufactured in small sizes which are desirable for portable instruments.

Coil Assembly

The moving coil is mounted on a light aluminum frame which is fitted around a soft-iron core. The soft-iron core offers little opposition to the magnetic lines of force emanating from the pole pieces (low reluctance). Therefore the purpose of the *soft-iron core* is to further *concentrate* the magnetic lines of force produced by the horseshoe magnet in the space occupied by the moving coil. The principal function of the *aluminum frame* is to *electrically dampen* the meter movement. From our discussion of the basic meter movement, we know that a magnetic field is established about the coil when there is current through it. The interaction of the magnetic fields due to the permanent magnet and current-carrying coil produces a torque which rotates the coil and pointer. Since the coil is mounted on the aluminum frame a voltage is induced in the aluminum frame. This induced voltage causes an induced current to flow in the aluminum frame, which results in a magnetic field that according to Lenz's law must oppose the motion that induced it. Thus an *electrical braking action results* which serves to dampen the meter movement. By carefully designing the coil and frame the desired damping characteristics can be obtained.

Spiral Springs

The two spiral springs (upper and lower) are used to supply current to the meter movement. In addition they supply a restoring torque to bring the coil and pointer to rest at the appropriate position. Normally the end of one spring is connected to a *zero adjust screw* on the face of the meter movement. This adjustment permits one to adjust the pointer to the zero position on the calibrated scale when no current is flowing through the meter movement. To a large extent the accuracy of the meter movement depends on the properties of the springs and their ability to remain constant with time, temperature, current, and so on. D'Arsonval's original movement had the zero position at midscale. By positioning the two spiral springs so that they tighten and relax together, the zero position can be moved from midscale to the extreme left.[3] The advantage of this arrangement is to increase the number of scale divisions between 0 and I_{FS}. A disadvantage is that the test leads must be connected with the proper polarity to avoid having the pointer attempt to deflect past zero. In Fig. 2-4 two retaining pins are employed to restrict the movement of the pointer so that it cannot move past the zero or full-scale positions. In practice both meter movements with the zero position at midscale and on the extreme left are encountered.

Suspension System—Jeweled Bearings

The moving coil, pointer, and associated components are mounted on a common shaft as illustrated in Fig. 2-4. This system is suspended via a *pivot-and-jewel* arrangement similar to the type employed in mechanical watches. A simplified diagram of this arrangement is illustrated in Fig. 2-5. The pivot is generally made from a steel alloy to obtain the proper hardness. Today the "jewel" is normally glass. The term "jewel" probably is a carry-over from the time when sapphire was employed.[4] The purpose of the suspension system is to keep friction to a minimum so that the coil and pointer are free to rotate. Notice in Fig. 2-4 that *counterweights* have been attached to

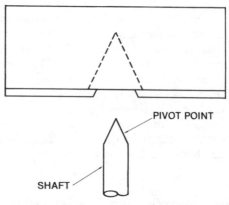

Fig. 2-5. Simplified jewel bearing.

the pointer so that a near-perfect balance of the pointer assembly is achieved (which decreases bearing friction).

2-6 TAUT-BAND AND OTHER MOVEMENTS

A significant variation of the d'Arsonval meter movement is the taut-band arrangement illustrated in Fig. 2-6. Notice that the spiral springs, jewel bearings, and pivots of the d'Arsonval movement have been replaced with two stretched (taut) metal bands. Spiral springs are not required because the metal bands twist as the coil rotates and thus provide the necessary restoring torque. Since no pivot-and-jewel arrangement is employed, in the taut-band movement, friction is greatly reduced. Thus the taut-band movement is generally *more sensitive* than a true d'Arsonval meter movement. In addition, taut-band movements are *more* rugged than their d'Arsonval counterparts, since there are no jewels to fracture should the movement be dropped (heaven forbid). In recent years taut-band movements

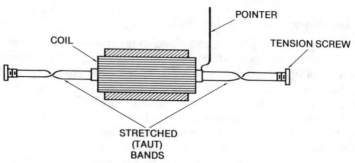

Fig. 2-6. Simplified taut-band movement.

have become increasingly popular due to the advantages cited above.

Both the taut-band and d'Arsonval meter movements provide a deflection which is directly proportional (approximately) to the *average* current through them. In later chapters you will learn how to modify these basic meter movements so that larger currents than I_{FS} can be measured. In addition you will learn how the basic movements can be used to measure dc voltages, ac voltages, resistance, and other quantities. There are many types of meter movements available, but the taut-band and d'Arsonval movements are by far the most popular. The symbol we will use to represent a basic meter movement is illustrated in Fig. 2-7. Recall that R_m is the internal resistance of the meter movement—this is essentially the resistance of the moving coil. In general, the smaller R_m is, the better. The current required to produce full-scale deflection (I_{FS}) is a good

Fig. 2-7. Symbol used to represent a basic meter movement.

$$R_m$$

indicator of the meter movement's sensitivity. You must not measure R_m directly with an ohmmeter; this could damage or destroy the meter movement! Why this is so will become clear when we study ohmmeters. Before leaving this section we will briefly discuss several other meter movements. Such meter movements are generally used in "special" applications.

Fig. 2-8 illustrates an *iron-vane* movement. In this movement two iron vanes (one fixed and one movable) are placed inside a stationary coil. When current flows through the stationary coil the vanes become magnetized and repel each other. Since the moving vane is attached to the pointer the pointer will deflect when current flows through the stationary coil. Notice that this movement is *mechanically damped* via a light aluminum damping vane. Iron-vane movements are not as sensitive as the d'Arsonval or taut-band movements discussed earlier. Iron-vane movements are primarily used for low-frequency ac measurements where rugged construction is desirable. Such movements are normally calibrated for the particular frequency they are designed to measure.

Another type of meter movement you may encounter is the *electrodynamometer* or simply *dynamometer*. A simplified sketch of this meter movement is provided in Fig. 2-9. Notice that two *fixed field coils* are used to produce a magnetic field. In the d'Arsonval and taut-band movements this was accomplished with a permanent horseshoe magnet. The two fixed coils are connected in series with a third movable coil which shares a common shaft with the pointer. When cur-

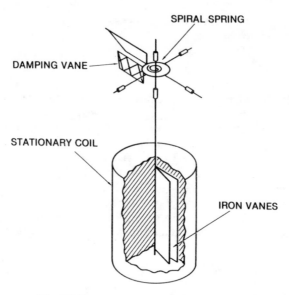

Fig. 2-8. Iron-vane movement (radial vane).

rent flows through the coils the magnetic fields interact to produce a torque which deflects the movable coil and pointer. An advantage of this type of movement is that it *responds equally well to ac signals and dc signals*. When we discuss rms measurements in a later chapter we will see the value of the electrodynamometer movement for calibration purposes. Disadvantages of the electrodynamometer movement include high power consumption, lower sensitivity than a d'Arsonval or taut-band movement, and a somewhat sluggish response. The electrodynamometer movement is often employed in audio circuits where reasonable amounts of power are available to drive the movement. Often the electrodynamometer movement is mechanically damped with vanes attached to the bottom of the shaft.

2-7 REVIEW OF OBJECTIVES

Real meter movements can only approximate the properties associated with ideal meter movements. The value I_{FS} is a good indication of a meter movement's sensitivity. The meter movement resistance, R_m, can be determined by the method illustrated in Experiment 2-1. It should *not* be measured directly with an ohmmeter because the meter movement is likely to be damaged due to excessive current from the ohmmeter. When you select a meter movement for a particular application special attention should be given to the spec-

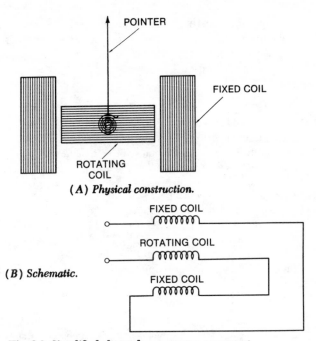

(A) *Physical construction.*

(B) *Schematic.*

Fig. 2-9. Simplified electrodynamometer movement.

ifications for input resistance, accuracy, sensitivity, and damping characteristics. Since accuracy is stated as a percentage of full-scale deflection the effective error is *not* a fixed percentage of the measured value. You should have a good understanding of the construction details of the d'Arsonval and taut-band movements. In addition you should be familiar with the principles which govern the iron-vane and electrodynamometer movements.

The following guidelines are useful in selecting and using a meter movement.

1. The smaller R_m, the better.
2. The smaller the stated accuracy as a percentage of the full-scale value, the better. Remember accuracy is stated in terms of error.
3. To minimize the effects of calibration error select a movement that produces up-scale readings for the range of currents you expect to measure.
4. The smaller is the value of I_{FS}, the more sensitive a meter movement is.
5. A meter movement should be critically damped or slightly underdamped so that the pointer will quickly come to rest on the measured value.

41

6. Always observe the polarity of the test leads when making a measurement. The red lead is positive (+) and the black lead negative (−) by convention.
7. Use common sense.

2-8 QUESTIONS

1. Why is a small value for R_m desirable?
2. What is calibration error?
3. How can the effects of calibration error be minimized?
4. Which is more sensitive, a 10-μA movement or a 50-μA movement?
5. What is the difference between responses that are overdamped, underdamped, and critically damped?
6. How can the period of a meter movement be decreased?
7. What is RCDRX?
8. List the major parts and functions of a d'Arsonval movement.
9. How does a taut-band movement differ from a d'Arsonval movement?
10. Why is a taut-band movement more sensitive than a d'Arsonval movement?

2-9 PROBLEMS

1. A meter movement has a full-scale current of 1 mA. The accuracy for this meter movement is stated as 2 percent of the full-scale value. Determine:
   (a) The amount of calibration error.
   (b) The range in actual current for measured currents of 0.9 mA and 0.4 mA.
   (c) Error as a percentage of the measured currents in (b).
2. A 50-μA meter movement deflects full scale when 250 mV is placed across the terminals of the meter movement. Estimate the internal resistance of the meter movement.
3. Dr. Courtine recommends the following changes in the construction of a d'Arsonval movement in order to reduce the cost of the instrument:
   (a) Replacing the aluminum frame with a wood frame.
   (b) Eliminating one sprial spring.
   (c) Utilizing a horseshoe magnet with flush pole pieces.
   (d) Replacing the soft-iron core with a hard-iron core.
   In addition Dr Courtine recommends replacing the aluminum bands in a taut-band movement with less expensive rubber bands. Comment on each of the proposed changes a–d relative to the performance of the modified instrument.

2-10 INTRODUCTION TO THE EXPERIMENTS

At the end of each chapter that follows you will find one or more experiments. The experiments are designed to supplement and complement the material presented in the text. The format utilized is as follows:

Objective—States the purpose of the experiment.

Material required—Specifies what is needed to do the experiment.

Introduction—Provides a background for the experiment.

Procedure—Specifies the steps to be followed in performing the experiment.

Discussion—(If necessary.) Reviews experimental results, provides suggestions for improving the experiment.

Conclusion—You should briefly discuss the experimental results here. Often questions are posed to guide your thinking. Ask yourself: What *concepts* are being demonstrated?

Many of the experiments specify a 50-μA meter movement. If you have a vom (utilizing a 50-μA movement) which permits direct access to its meter movement (most do), you can use the vom in this mode for most of the experiments. The author has obtained excellent results using a Simpson 260® vom on its 50-μA range for this purpose. If you do not have a 50-μA meter movement, other meter movements can be employed *if* you make appropriate changes in the experiment. For example, in Experiment 3-2 an ammeter is designed that provides ranges of 250 μA, ($5I_{FS}$), 500 μA ($10I_{FS}$), and 1 mA ($20I_{FS}$). If you had a 1-mA meter movement instead of a 50-μA meter movement, you could elect to design the ammeter to provide ranges of 5 mA, 10 mA, and 20 mA, respectively. Although many different meter movements can be employed in the experiments the author recommends a 50-μA movement if at all possible. If a 50-μA movement is used, no modifications in the experiments will be required.

2-11 EXPERIMENT 2-1

Objectives

The objective of this experiment is to determine the internal resistance (R_m) of a meter movement.

Material Required

50-μA meter movement (d'Arsonval or taut-band)
9-V battery (other voltages than 9 V can be used if desired)
Resistor decades: 1-kΩ, 100-Ω, and 10-Ω steps (although less convenient potentiometers and fixed resistors can be used if resistor decades are not available)

Introduction

The internal resistance (R_m) of a meter movement is essentially the resistance of the moving coil employed in the meter movement. You should *not* attempt to measure this resistance directly with an ohmmeter because most ohmmeters will send a current through the

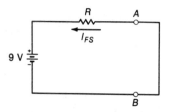

Fig. 2-10. Original circuit of
Experiment 2-1.

meter movement which is greater than the value required for full-scale deflection (I_{FS})! You wish to devise a method for determining R_m which is simple, reasonably accurate, and of course doesn't damage the meter movement. Your starting point is knowing how much current produces full-scale deflection (I_{FS}). This is easily determined since it is indicated on the face of the meter movement! Next you select a "convenient" voltage source. Since 9-V batteries are readily available and reasonably inexpensive we have selected one of them as our voltage source. Consider the circuit in Fig. 2-10. From Ohm's law it is obvious that the value of R required to produce a current equal to I_{FS} is 9 V/I_{FS}. In Fig. 2-11 we have replaced the conductor between A and B with our meter movement. The current

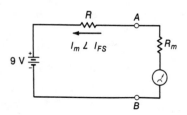

Fig. 2-11. Inserting the
meter movement.

that flows through the meter movement (I_m) will be less than I_{FS} because the resistance of the meter movement (R_m) has been added to the circuit. Thus

$$I_m = \frac{9 \text{ V}}{R + R_m}$$

which is less than I_{FS} since

$$I_{FS} = \frac{9 \text{ V}}{R}$$

from Fig. 2-10.

We can, however, adjust R so that the current through the meter (I_m) equals the full-scale value (I_{FS}). Once this is done we add a resistor (R_2) in parallel with the meter movement (Fig. 2-12). Resistor R_2 is adjusted until the meter movement indicates half-scale

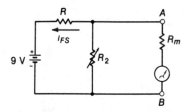

Fig. 2-12. Adjusting R_2 for
half-scale deflection.

deflection. Since the total current I_{FS} divides equally between R_2 and R_m it follows that R_2 must be equal to R_m.

Procedure

Step 1. Record the value of I_{FS} for your meter movement.

$$I_{FS} = \text{\underline{\hspace{3cm}}}$$

Step 2. Select a convenient voltage source (V). Calculate the value of R required to produce a current equal to I_{FS}.

$$R = \frac{V}{I_{FS}} = \text{\underline{\hspace{3cm}}}$$

Step 3. To compensate for resistor tolerances, variations in battery voltage, etc., adjust the decade used for R to a value 10 percent *higher* than the value calculated in Step 2.

Step 4. Build the circuit shown in Fig. 2-11. Adjust R to get exactly full-scale deflection.

Step 5. Connect the second decade (R_2) in parallel with the meter movement (Fig. 2-12). Initially set R_2 to 10 kΩ.

Step 6. Decrease the value of R_2 until the current through the meter movement (I_m) is one-half of I_{FS}. When this occurs $R_2 = R_m$. Record the value of R_2; this is the internal resistance of your meter.

$$R_2 = \text{\underline{\hspace{3cm}}}$$

Conclusion

In your own words summarize the results of the experiment. What are the significant sources of error in this experiment? How could significant sources of error be minimized?

2-12 REFERENCES

1. Leeds and Northrup Company, "Fundamentals of D-C Electrical Measurements," *Supplementary Information A0.0001*, North Wales, PA, p. 3.
2. Ibid., p. 3.
3. Houglum, Roger J. 1980. *Electronics Concepts, Applications, and History.* Boston: Brenton Publishers, p. 124.
4. Cunningham, John. 1973. *Understanding and Using the VOM and EVM.* Blue Ridge Summit, PA: TAB Books, p. 10.

CHAPTER 3

DC Ammeters

3-1 INTRODUCTION

In this chapter you will learn how to increase the range of a basic meter movement by employing one or more *shunt* resistors. This will enable you to measure currents larger than I_{FS} (the full-scale deflection meter current). Since the Ayrton shunt ammeter is found in many commercial instruments you will study it in detail. In addition, you will learn how to estimate the accuracy of dc current measurements and the amount of loading (error) produced by placing the measurement instrument in the circuit.

3-2 OBJECTIVES

At the end of this chapter you will be able to do the following:

- Determine the value of the shunt resistance required to increase the current measuring range of the basic meter movement.
- Draw an equivalent circuit for the shunted movement.
- Analyze and design elementary dc ammeters.
- Analyze and design Ayrton shunt ammeters.
- Estimate the accuracy and loading error for a particular dc current measurement.
- Specify the conditions necessary to minimize ammeter loading error.

3-3 ELEMENTARY AMMETERS: SHUNTED MOVEMENTS

The basic meter movements discussed in the previous chapter can be used to measure currents up to the full-scale value (I_{FS}). If you wish to measure currents larger than I_{FS} (you do!) a *shunt resistor* (R_{sh}) is employed in parallel with the meter movement. Such an arrangement is illustrated in Fig. 3-1. We wish to determine the

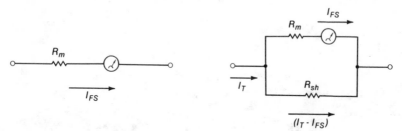

(A) *Basic meter movement measures currents up to I_{FS}.*

(B) *Shunted meter movement measures currents up to I_T.*

Fig. 3-1. Shunting a meter movement.

value of R_{sh} so that when I_T (total current to be measured) enters the shunted meter movement, a full-scale deflection of the basic meter movement results. Note, that in Fig. 3-1, the current I_T divides between R_m and R_{sh} so that I_{FS} flows through R_m and $I_T - I_{FS}$ flows through R_{sh}. Thus

$$I_{sh} = I_T - I_{FS}$$

$$V_{Rsh} = V_{Rm} = I_{FS} R_m$$

Therefore

$$R_{sh} = \frac{V_{Rsh}}{I_{Rsh}} = \frac{I_{FS} R_m}{I_T - I_{FS}} \tag{3-1}$$

The *equivalent resistance* of the shunted meter movement (R_m') is

$$R_m' = R_{sh} \parallel R_m = \frac{R_{sh} R_m}{R_{sh} + R_m} \tag{3-2}$$

The symbol $\parallel$ is pronounced, "in parallel with." Note that the shunted meter movement acts like a basic meter movement whose $I_{FS} = I_T$ and whose $R_m = R_m'$. By employing a shunt resistance you effectively *increase the range* of the basic meter movement. The price you pay for doing this is to *decrease the sensitivity* of the instrument, since a current equal to I_T is now required for full-scale deflection.

EXAMPLE 3-1

A basic meter movement has an $I_{FS} = 50$ μA and $R_m = 2$ kΩ. Determine: (a) the shunt resistance required so that currents up to 250 μA (I_T) can be measured, (b) the equivalent resistance of the shunted meter movement, (c) an equivalent circuit for the shunted meter movement.

(a)

$$R_{sh} = \frac{I_{FS}R_m}{I_T - I_{FS}} = \frac{(50\ \mu\text{A})\ (2\ \text{k}\Omega)}{250\ \mu\text{A} - 50\ \mu\text{A}} = \frac{100\ \text{mV}}{200\ \mu\text{A}} = 500\ \Omega$$

(b)

$$R_m{}' = R_{sh}\ ||\ R_m = \frac{(2\ \text{k}\Omega)\ (0.5\ \text{k}\Omega)}{2.5\ \text{k}\Omega} = 400\ \Omega$$

(c) The shunted meter movement acts like a basic meter movement whose $I_{FS} = I_T = 250$ μA and whose $R_m = R_m{}' = 400$ Ω. It and its equivalent circuit are illustrated in Fig. 3-2.

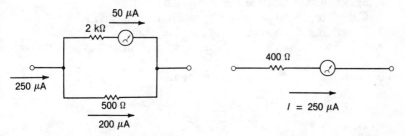

(A) Shunted meter movement.   (B) Equivalent circuit.

Fig. 3-2. Equivalent circuit for Example 3-1.

With the 50-μA movement employed in Example 3-1, it is possible to measure currents larger than 250 μA if shunt resistors smaller than 500 Ω are employed. From a practical point of view there will be a limit on how small R_{sh} can be made. The following example illustrates this fact.

EXAMPLE 3-2

Using the 50-μA, 2-kΩ movement employed in Example 3-1, determine R_{sh} for the following ranges: (a) 100 mA, (b) 1 A, (c) 10 A.

(a)

$$I_T = 100\ \text{mA} = 100\ 000\ \mu\text{A}$$

$$R_{sh} = \frac{I_{FS}R_m}{I_T - I_{FS}} = \frac{(50\ \mu\text{A})\ (2\ \text{k}\Omega)}{99\ 950\ \mu\text{A}} = 1\ \Omega$$

Similarly,

(b)

$$R_{sh} = \frac{(50 \ \mu A) \ (2 \ k\Omega)}{999 \ 950 \ \mu A} = 100 \ m\Omega$$

(c)

$$R_{sh} = \frac{(50 \ \mu A) \ (2 \ k\Omega)}{9 \ 999 \ 950 \ \mu A} = 10 \ m\Omega$$

Thus, when the value required for R_{sh} is unreasonably small (perhaps on the same order of magnitude as a short length of wire), you might consider employing a *less* sensitive basic meter movement and/or a different measurement technique.

3-4 DESIGNING AN ELEMENTARY AMMETER

You can design an elementary ammeter using a commercially available meter movement by following the procedure given below:

1. Select an appropriate meter movement.
2. Calculate the required shunt resistance for each range.
3. Determine the power rating for each shunt resistance.
4. Select a suitable switching arrangement for the different shunts.
5. Build and calibrate the instrument.

The design process is illustrated in Example 3-3.

EXAMPLE 3-3

Assuming that a 50-μA, 2-kΩ movement is available, design an ammeter that provides the following ranges: 0–1 mA, 0–10 mA, and 0–100 mA.

Being rather astute, you have undoubtedly noted that the shunt resistance required for the 100-mA range was calculated in the previous example problem! Thus, for the 100-mA range, $R_{sh} = 1 \ \Omega$.

Now for the other ranges:

1-mA range:

$$R_{sh} = \frac{I_{FS}R_m}{I_T - I_{FS}} = \frac{(50 \ \mu A) \ (2 \ k\Omega)}{950 \ \mu A} = 105.3 \ \Omega$$

10-mA range:

$$R_{sh} = \frac{I_{FS}R_m}{I_T - I_{FS}} = \frac{(50 \ \mu A) \ (2 \ k\Omega)}{9 \ 950 \ \mu A} = 10.05 \ \Omega$$

Knowing the value of R_{sh} and I_{sh} you can easily calculate the maximum power each shunt resistance will be required to dissipate from $P = I_{sh}^2 R_{sh}$. Thus:

1-mA range:

$$P = (0.95 \text{ mA})^2(105.3 \ \Omega) = 0.095 \ 03 \text{ mW}$$

10-mA range:

$$P = (9.95 \text{ mA})^2(10.05 \ \Omega) = 0.995 \text{ mW}$$

100-mA range:

$$P = (99.95 \text{ mA})^2(1 \ \Omega) = 9.99 \text{ mW}$$

Obviously there is no problem obtaining commercial resistors that can handle the powers calculated above. The input resistance (R_{in}) for each range R_m' is $R_{sh} \parallel R_m$. This information is provided in Table 3-1. With the addition of a simple rotary switch the final design is illustrated in Fig. 3-3.

Table 3-1. Input Resistance for the Ammeter in Example 3-3

Range	$R_{\text{in}} = R_m' = R_{sh} \parallel R_m$
1 mA	$105.3 \ \Omega \parallel 2 \text{ k}\Omega \cong 100 \ \Omega$
10 mA	$10.05 \ \Omega \parallel 2 \text{ k}\Omega \cong 10 \ \Omega$
100 mA	$1 \ \ \ \Omega \parallel 2 \text{ k}\Omega \cong 1 \ \Omega$

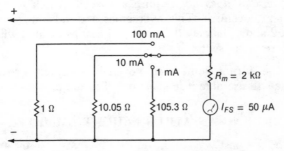

Fig. 3-3. Ammeter for Example 3-3.

The ammeter which we just designed has a significant drawback. To understand this drawback, assume that 0.8 mA is being measured on the 10-mA range. In order to reduce the effect of calibration error (and your eyestrain) you wisely decide to switch to the 1-mA range. During the time the switch moves from the 10-mA position to the 1-mA position *no* shunt resistance is connected across the meter movement! Such a state of affairs is most undesirable, and if you were not careful you could easily damage the meter movement. Some possible solutions to this problem include the following:

1. Provide instructions with the ammeter that indicate the instrument should be removed from the circuit under test prior to changing ranges. This solution is not very practical in light of our current perception of "human nature" and the obvious inconvenience associated with the measurement.

2. Employ a multiposition make-before-break switch in place of the simple rotary switch. As the name implies such a switch "makes" the new connection before "breaking" the old connection. This arrangement ensures that the meter movement will be protected with a combination of two shunt resistances when you switch ranges.
3. Utilize a universal, i.e., an *Ayrton* shunt, design for the ammeter. This important circuit is illustrated in Fig. 3-4. Note that

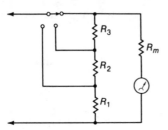

Fig. 3-4. Ayrton shunt ammeter.

this circuit *does not* require a special switch to protect the meter movement. This overcomes a disadvantage of solution No. 2 above, namely the increased cost associated with a make-before-break switch.

Since the Ayrton shunt is employed in many commercial instruments we will examine it in some detail in the next section.

3-5 DESIGNING AN AYRTON SHUNT AMMETER

Perhaps the best way to introduce the Ayrton shunt ammeter illustrated in Fig. 3-4 is by "brute force." Thus, we anxiously proceed to Example 3-4.

EXAMPLE 3-4

Design an Ayrton shunt ammeter employing (can you guess?) a 50-μA, 2-kΩ movement, that provides the following ranges: 0–1 mA, 0–10 mA, and 0–100 mA.

In order to meet this "modest" objective you can employ the following procedure for the Ayrton shunt ammeter illustrated in Fig. 3-4.

1. Define the switch positions.
2. Identify the effective shunt resistance (R_{shE}) and the effective meter resistance (R_{mE}) for *each* switch position.
3. Apply Equation 3-1,

$$R_{sh} = \frac{I_{FS}R_m}{I_T - I_{FS}}$$

to each switch position, where $R_{sh} = R_{shE}$ and $R_m = R_{mE}$.

4. Take a refresher course in elementary algebra in order to solve the simultaneous equations generated in Step 3.

Our definitions of the switch positions and resulting effective shunt and meter resistances are illustrated in Fig. 3-5.

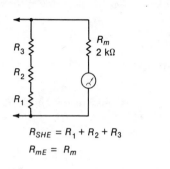

$$R_{SHE} = R_1 + R_2 + R_3$$
$$R_{mE} = R_m$$

(A) 1-mA position.

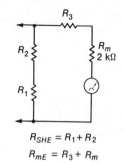

$$R_{SHE} = R_1 + R_2$$
$$R_{mE} = R_3 + R_m$$

(B) 10-mA position.

(C) 100-mA position.

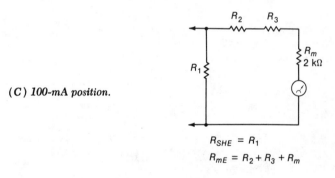

$$R_{SHE} = R_1$$
$$R_{mE} = R_2 + R_3 + R_m$$

Fig. 3-5. Switch positions for Example 3-4.

The author hopes that the procedure outlined above will prove to be less than traumatic on the part of the reader. In any event the details which follow should clarify the procedure. From Fig. 3-5:

Position 1 (1 mA):

$$R_{mE} = R_m = 2\,\text{k}\Omega$$
$$R_{shE} = R_1 + R_2 + R_3 = \frac{(50\,\mu\text{A})\,(2\,\text{k}\Omega)}{950\,\mu\text{A}} = 105.3\,\Omega$$

Position 2 (10 mA):

$$R_{mE} = R_m + R_3 = 2\,\text{k}\Omega + R_3$$
$$R_{shE} = R_1 + R_2 = \frac{(50\,\mu\text{A})\,(2\,\text{k}\Omega + R_3)}{9\,950\,\mu\text{A}}$$

Position 3 (100 mA):

$$R_{mE} = R_m + R_2 + R_3 = 2 \text{ k}\Omega + R_2 + R_3$$
$$R_{shE} = R_1 = \frac{(50 \ \mu\text{A})\,(2 \text{ k}\Omega + R_2 + R_3)}{99\ 950 \ \mu\text{A}}$$

The desired simultaneous equations are summarized as follows:

$$R_1 + R_2 + R_3 = 105.3 \ \Omega \tag{1}$$

$$R_1 + R_2 = \frac{50(2 \text{ k}\Omega + R_3)}{9\ 950} \tag{2}$$

$$R_1 = \frac{50(2 \text{ k}\Omega + R_2 + R_3)}{99\ 950} \tag{3}$$

There are of course numerous exciting methods available to solve such a set of equations. For this example we elect the method of substitution. Thus

$$R_1 + R_2 = 105.3 \ \Omega - R_3$$

from (1). Substituting into (2) yields

$$105.3 \ \Omega - R_3 = \frac{100 \times 10^3 \ \Omega + 50R_3}{9.95 \times 10^3}$$

$$(105.3 \ \Omega - R_3)\,(9.95 \times 10^3) = 100 \times 10^3 \ \Omega + 50R_3$$

$$1\ 047.74 \times 10^3 \ \Omega - 9.95 \times 10^3 R_3 = 100 \times 10^3 \ \Omega + 50R_3$$

$$1\ 047.74 \times 10^3 \ \Omega - 100 \times 10^3 \ \Omega = 9.95 \times 10^3 R_3 + 50R_3$$

$$947.74 \times 10^3 \ \Omega = 10 \times 10^3 R_3$$

$$R_3 = \frac{947.74 \times 10^3}{10 \times 10^3} = 94.774 \cong 94.8 \ \Omega$$

Now, from (1),

$$R_1 = 105.3 \ \Omega - (R_2 + R_3)$$

Substituting into (3) yields

$$[105.3 \ \Omega - (R_2 + R_3)] = \frac{100 \times 10^3 \ \Omega + 50(R_2 + R_3)}{99.95 \times 10^3}$$

$$[105.3 \ \Omega - (R_2 + R_3)]\,99.95 \times 10^3 = 100 \times 10^3 \ \Omega + 50(R_2 + R_3)$$

$$10\ 524.7 \times 10^3 \ \Omega - 99.95 \times 10^3(R_2 + R_3) = $$
$$100 \times 10^3 \ \Omega + 50(R_2 + R_3)$$

$$10\ 524.7 \times 10^3 \ \Omega - 100 \times 10^3 \ \Omega = $$
$$50(R_2 + R_3) + 99.95 \times 10^3(R_2 + R_3)$$

$$10\ 424.7 \times 10^3 \ \Omega = 100 \times 10^3(R_2 + R_3)$$

$$R_2 + R_3 = \frac{10\ 424.7 \times 10^3}{100 \times 10^3} = 104.247 \cong 104.25 \ \Omega$$

Finally,

$$R_2 + R_3 = 104.25 \ \Omega$$
$$R_2 + 94.8 \ \Omega = 104.25 \ \Omega$$
$$R_2 = 104.25 - 94.8 = 9.45 \ \Omega$$

Thus

$$R_1 + R_2 + R_3 = 105.3\ \Omega$$
$$R_1 + 9.45\ \Omega + 94.8\ \Omega = 105.3\ \Omega$$
$$R_1 = 105.3 - 9.45 - 94.8 = 1.05\ \Omega$$

We proudly present the final design in Fig. 3-6. The input resistance (R_{in}) for each range is $R_{shE}\ ||\ R_{mE}$. This information is provided

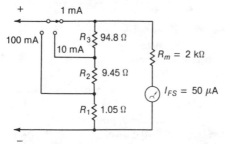

Fig. 3-6. Ayrton shunt ammeter for Example 3-4.

in Table 3-2. Note that the input resistance for each range in Table 3-2 is only *slightly larger* than the corresponding values for Example 3-3 (Table 3-1).

Table 3-2. Input Resistance for the Ayrton Shunt Ammeter in Example 3-4

Range	$R_{in} = R_{shE}\ ||\ R_{mE}$		
1 mA	105.3 Ω	2 kΩ $\cong$	100 Ω
10 mA	10.5 Ω	2.094 8 kΩ $\cong$	10.45 Ω
100 mA	1.05 Ω	2.104 3 kΩ $\cong$	1.05 Ω

Before concluding the discussion of the Ayrton shunt ammeter, we will check the values calculated for R_1, R_2, and R_3. You can do this by employing the *current division* principle to verify that full-scale deflection occurs, for each switch position, when I_T enters the ammeter. Readers not familiar with the current division principle are referred to Appendix C. Thus for any switch position

$$I_m = \frac{I_T\,R_{shE}}{R_{shE} + R_{mE}} \qquad (3\text{-}3)$$

For the 1-mA position

$$I_m = \frac{(1\ \text{mA})\,(105.3\ \Omega)}{2\ 105.3\ \Omega} \cong 50\ \mu\text{A}$$

For the 10-mA position

$$I_m = \frac{(10\ \text{mA})\,(10.5\ \Omega)}{2\ 105.3\ \Omega} \cong 50\ \mu\text{A}$$

For the 100-mA position

$$I_m = \frac{(100\ mA)\ (1.05\ \Omega)}{2\ 105.3\ \Omega} \cong 50\ \mu A$$

Neglecting rounding errors the answers "check" and we breathe a collective sigh of relief.

3-6 AMMETER LOADING EFFECTS

You are now in the enviable position of being able to modify a basic meter movement so that a wide range of currents can be measured. How "good" are the instruments designed in the previous sections? The answer to this question is applicable to all instruments. *They are as good as your understanding of their limitations and the circuits in which they are used!* The purpose of this section is to clarify some of these limitations. As you learned in Chapter 1, whenever an instrument is used to make a measurement, the system being measured is changed by the presence of the instrument. If only "small" changes occur, then the measured value will closely approximate the original value that is the value of the unknown before the instrument was placed in the system. In this case the measured value provides an *accurate* description of the original system. If on the other hand "large" changes are introduced by the instrument, then the measured value will differ significantly from the original value of the unknown. Thus the *error* associated with the measurement is substantial. Obviously, you want your measurements to be as accurate as possible.

Consider the current measurement illustrated in Fig. 3-7. We wish to measure the current in branch *AB*. The internal resistance of the

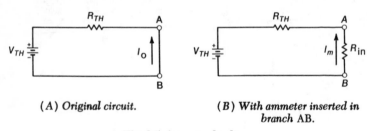

(A) *Original circuit.* (B) *With ammeter inserted in branch* AB.

Fig. 3-7. Ammeter loading.

ammeter (R_{in}) is added to branch *AB* when the ammeter is placed in the circuit. You can think of the original circuit as a very simple circuit, or the Thevenin equivalent of a more complex circuit. We will consider the original circuit in Fig. 3-7 to be the Thevenin equivalent of a more complex circuit. This will enable us to obtain very general results which will be applicable to virtually any circuit in which we desire to measure current. Readers who are not familiar with Thevenin's theorem are referred to Appendix E.

From Fig. 3-7A the *original* current (I_o) is easily determined with Ohm's law:

$$I_o = \frac{V_{TH}}{R_{TH}}$$

where

V_{TH} = Thevenin voltage,
R_{TH} = Thevenin resistance.

Similarly, the *measured* current (I_m) in Fig. 3-7B is

$$I_m = \frac{V_{TH}}{R_{TH} + R_{in}}$$

where R_{in} is the ammeter's internal resistance.

The *accuracy* (a) of the current measurement depends on how close the measured current (I_m) is to the original current (I_o). Thus we can define the accuracy a as

$$a = \frac{I_m}{I_o} \tag{3-4}$$

Therefore

$$a = \frac{V_{TH}}{R_{TH} + R_{in}} \div \frac{V_{TH}}{R_{TH}} = \frac{V_{TH}}{R_{TH} + R_{in}} \times \frac{R_{TH}}{V_{TH}}$$

$$= \frac{R_{TH}}{R_{TH} + R_{in}} \tag{3-5}$$

where R_{TH} and R_{in} are as previously defined.

The *error* (e) associated with the measurement is

$$e = \frac{I_o - I_m}{I_o} \tag{3-6}$$

where

I_o is the original current,
I_m is the measured current.

You can think of the error (e) as the amount of "inaccuracy" associated with the measurement. This type of error occurs because the ammeter requires power from the circuit to deflect the moving coil in the meter movement. The ammeter is said to *load* the circuit. This is why this type of error is called *loading* error. Since loading error can be thought of as "*inaccuracy*" we can express it as follows:

$$e = 1 - a \tag{3-7}$$

where a is the accuracy as previously defined.

The utility of the relationships given above and some additional details are illustrated in the next few examples.

EXAMPLE 3-5

For the circuits illustrated in Fig. 3-8 determine the following: (a) The original current I_o, (b) the measured current (I_m), (c)

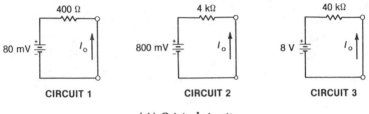

(A) Original circuits.

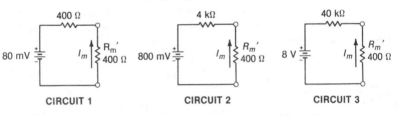

(B) Ammeter of Example 3-1 ($R_m' = 400 \ \Omega$) inserted to measure current.

Fig. 3-8. Circuits for Example 3-5.

the percent of accuracy, and (d) the percent of error. Assume that loading error is the only significant error and that the ammeter designed in Example 3-1 ($I_{FS} = 250 \ \mu A$, $R_m' = R_{in} = 400 \ \Omega$) is used for each measurement.

Circuit 1:

$$I_o = \frac{80 \text{ mV}}{400 \ \Omega} = 200 \ \mu A$$

$$a = \frac{R_{TH}}{R_{TH} + R_m} = \frac{400 \ \Omega}{400 \ \Omega + 400 \ \Omega} = 0.5 = 50\%$$

$$I_m = aI_o$$
$$= 0.5(200 \ \mu A) = 100 \ \mu A$$
$$e = 1 - a = 1 - 0.5 = 0.5 = 50\%$$

Circuit 2:

$$I_o = \frac{800 \text{ mV}}{4 \text{ k}\Omega} = 200 \ \mu A$$

$$a = \frac{R_{TH}}{R_{TH} + R_m} = \frac{4 \text{ k}\Omega}{4 \text{ k}\Omega + 400 \ \Omega} = 0.909 = 90.9\%$$

$$I_m = aI_o$$
$$= 0.909(200 \ \mu A) = 181.8 \ \mu A$$
$$e = 1 - a = 1 - 0.909 = 9.1\%$$

Circuit 3:

$$I_o = \frac{8 \ V}{40 \ k\Omega} = 200 \ \mu A$$

$$a = \frac{R_{TH}}{R_{TH} + R_m} = \frac{40 \ k\Omega}{40 \ k\Omega + 400 \ \Omega} = 0.99 = 99\%$$

$$I_m = aI_o = 0.99(200 \ \mu A) = 198 \ \mu A$$

$$e = 1 - a = 1 - 0.99 = 0.01 = 1\%$$

In these calculations the accuracy (a) was calculated first, followed by the measured current (I_m). If you prefer, this procedure can be reversed. For example, in circuit 1,

$$I = I_o = \frac{80 \ mV}{400 \ \Omega} = 200 \ \mu A$$

$$I_m = \frac{80 \ mV}{400 \ \Omega + 400 \ \Omega} = 100 \ \mu A$$

$$a = \frac{I_m}{I_o} = \frac{100 \ \mu A}{200 \ \mu A} = 0.5 = 50\%$$

$$e = 1 - a = 1 - 0.5 = 0.5 = 50\%$$

The results of these calculations are summarized in Table 3-3.

Table 3-3. Summary of Calculations for Example 3-5

Circuit	I_o	I_m	a	e
1	200 μA	100 μA	50 %	50 %
2	200 μA	181.8 μA	90.9%	9.1%
3	200 μA	198 μA	99 %	1 %

You should note that "good" accuracy results when $R_{TH} >> R_m$. What is considered "good" is somewhat subjective. Frequently the definition of good depends on the specific application, peer pressure, the amount of money available to purchase an instrument, etc. For our purposes we will define accuracy to be good if the error $e \leqq 10$ percent. Thus

$$a = \frac{R_{TH}}{R_{TH} + R_m}$$
$$a(R_{TH} + R_m) = R_{TH}$$
$$aR_{TH} + aR_m = R_{TH}$$
$$R_{TH} - aR_{TH} = aR_m$$
$$R_{TH}(1-a) = aR_m$$

$$R_{TH} = \frac{aR_m}{1 - a}$$

Substituting e for $1 - a$ yields

$$R_{TH} = \frac{aR_m}{e} \qquad (3\text{-}8)$$

as the Thevenin resistance.

Equation 3-8 enables you to construct Table 3-4, which indicates the relative size of R_{TH} for good accuracy. You should note that Equation 3-8 and Table 3-4 deal *only* with the error that results from the insertion of the ammeter into the circuit (loading error). Other sources of error are *not* included since loading error is often the most significant error. The results summarized in Table 3-4 are quite useful, because you can use it to predict the amount of error in a current measurement, based on the resistance of the ammeter and the resistance of the circuit.

Table 3-4. Relative Size of R_{TH} for Good Ammeter Accuracy

Percent Accuracy ($a_i \times 100$)	Percent Error ($e \times 100$)	$R_{TH} = \dfrac{a_i R_m}{e}$
99	1	$99R_m$
98	2	$49R_m$
97	3	$32.3R_m$
96	4	$24R_m$
95	5	$19R_m$
94	6	$15.7R_m$
93	7	$13.3R_m$
92	8	$11.5R_m$
91	9	$10.1R_m$
90	10	$9R_m$

EXAMPLE 3-6

Determine what the ammeter of Example 3-3 will read if it is used to measure I in Fig. 3-9. Assume the ammeter is used on the 0- to 1-mA range ($R_m' \cong 100 \ \Omega$) and that loading error is the only significant error.

First calculate R_{TH} (see Fig. 3-10):

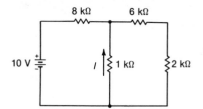

8 kΩ   6 kΩ

10 V I 1 kΩ   2 kΩ   **Fig. 3-9. Circuit for Example 3-6.**

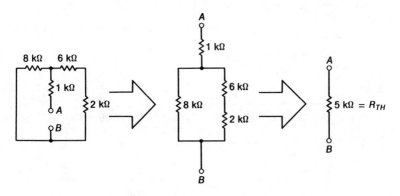

Fig. 3-10. Solving for R_{TH} in Example 3-6.

$$R_{TH} = R_{AB} = 8\text{ k}\Omega \,||\, (6\text{ k}\Omega + 2\text{ k}\Omega) + 1\text{ k}\Omega$$
$$= 8\text{ k}\Omega \,||\, 8\text{ k}\Omega + 1\text{ k}\Omega$$
$$= 4\text{ k}\Omega + 1\text{ k}\Omega$$
$$= 5\text{ k}\Omega$$

Next we calculate V_{TH}. Notice in Fig. 3-11 that when there is an open circuit between A and B the current through the 1-kΩ resistor

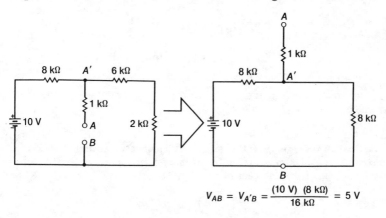

$$V_{AB} = V_{A'B} = \frac{(10\text{ V})\ (8\text{ k}\Omega)}{16\text{ k}\Omega} = 5\text{ V}$$

Fig. 3-11. Solving for V_{TH} in Example 3-6.

is zero. Thus the voltage across the 1-kΩ resistor is also zero and $V_{AB} = V_{A'B} = V_{TH}$. Therefore

$$V_{TH} = \frac{(10\text{ V})\ (8\text{ k}\Omega)}{16\text{ k}\Omega} = 5\text{ V}$$

When the ammeter is inserted between A and B it adds 100 Ω to the branch. This is illustrated in Fig. 3-12C. Thus

61

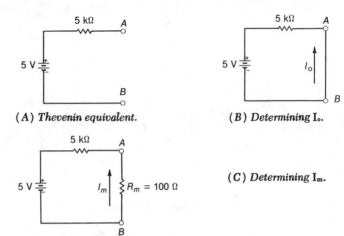

Fig. 3-12. Using the Thevenin equivalent circuit of Examples 3-6 to determine I_o and I_m.

$$I_m = \frac{V_{TH}}{R_{TH} + R_m} = \frac{5 \text{ V}}{5 \text{ k}\Omega + 100 \text{ }\Omega} = 0.98 \text{ mA}$$

or, if you prefer, from Fig. 3-12B,

$$I_o = \frac{V_{TH}}{R_{TH}} = \frac{5 \text{ V}}{5 \text{ k}\Omega} = 1 \text{ mA}$$

$$a = \frac{R_{TH}}{R_{TH} + R_m} = \frac{5 \text{ k}\Omega}{5 \text{ k}\Omega + 100 \text{ }\Omega} = 0.98$$

$$I_m = aI_o = 0.98(1 \text{ mA}) = 0.98 \text{ mA}$$

Note that this accuracy (98 percent) is in good agreement with Table 3-4, since R_m is 50 times R_{TH}.

Example 3-7

Predict the range of the measured current in Example 3-6, assuming the accuracy on the 1-mA range is 3 percent of the full-scale value. Three percent of 1 mA is

$$\frac{3}{100}(1 \text{ mA}) = 0.03 \text{ mA}$$

Thus the calibration error is 0.03 mA. The expected range R is $R = I_m \pm \text{CE}$:

$$R = (0.98 \pm 0.03)\text{mA}$$
$$= 0.95 \text{ mA to } 1.01 \text{ mA}$$

As circuit complexity increases we quickly reach a point where R_{TH} is not easily determined. In such a case how can we be sure that our measured values are accurate? This pragmatic question is addressed in the next example.

EXAMPLE 3-8

Will the measurement illustrated in Fig. 3-13 be accurate? Justify your answer and estimate the accuracy of the measurement.

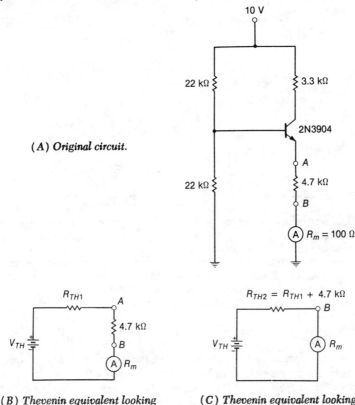

(A) Original circuit.

(B) Thevenin equivalent looking into A.

(C) Thevenin equivalent looking into B.

Fig. 3-13. Circuit and Thevenin equivalents for Example 3-8.

First, consider the Thevenin equivalent circuit from point A to ground. Perhaps we cannot "easily" determine the values of R_{TH} and V_{TH}, but we *can* easily picture the circuit shown in Fig. 3-13B. Next, consider the Thevenin equivalent circuit from point B to ground. This is illustrated in Fig. 3-13C. In this figure

$$a = \frac{R_{TH2}}{R_{TH2} + R_m} = \frac{R_{TH1} + 4.7 \text{ k}\Omega}{R_{TH1} + 4.7 \text{ k}\Omega + 100 \text{ }\Omega}$$

Even though you do not know the value of R_{TH1} you do know that one of the following three conditions must exist:

(1) $R_{TH1} \ll 4.7 \text{ k}\Omega$

(2) $R_{TH1} \cong 4.7 \text{ k}\Omega$

(3) $R_{TH1} \gg 4.7 \, \text{k}\Omega$

In order to arrive at viable "ball-park" figures, we will employ the 10:1 rule to define large and small.

Consider Case 1: $R_{TH1} \ll 4.7 \, \text{k}\Omega$.

$$R_{TH1} = \frac{4.7 \, \text{k}\Omega}{10} = 0.47 \, \text{k}\Omega$$

$$a = \frac{0.47 \, \text{k}\Omega + 4.7 \, \text{k}\Omega}{0.47 \, \text{k}\Omega + 4.7 \, \text{k}\Omega + 100 \, \Omega} = \frac{5.17 \, \text{k}\Omega}{5.27 \, \text{k}\Omega} = 98.1\%$$

Consider Case 2: $R_{TH1} \cong 4.7 \, \text{k}\Omega$.

$$a = \frac{4.7 \, \text{k}\Omega + 4.7 \, \text{k}\Omega}{4.7 \, \text{k}\Omega + 4.7 \, \text{k}\Omega + 100 \, \Omega} = \frac{9.4 \, \text{k}\Omega}{9.5 \, \text{k}\Omega} = 98.9\%$$

Consider Case 3: $R_{TH1} \gg 4.7 \, \text{k}\Omega$.

$$R_{TH1} = 10(4.7 \, \text{k}\Omega) = 47 \, \text{k}\Omega$$

$$a = \frac{47 \, \text{k}\Omega + 4.7 \, \text{k}\Omega}{47 \, \text{k}\Omega + 4.7 \, \text{k}\Omega + 100 \, \Omega} = \frac{51.7 \, \text{k}\Omega}{51.8 \, \text{k}\Omega} = 99.8\%$$

Thus, regardless of the value for R_{TH1} we can be confident that the measured value will be accurate. If R_{TH} is not easily determined, we are assured of good accuracy if $R \gg R_m$, where R is the resistance in *series* with the ammeter. Note that the previous calculations did not take into account calibration error. Remember, the way to minimize calibration error is to pick an instrument (or range) that provides "up-scale" readings.

3-7 REVIEW OF OBJECTIVES

Currents larger than I_{FS} can be measured by employing one or more shunt resistors in parallel with the basic meter movement. The effect of a shunt resistor is to increase the range and to decrease the sensitivity of the ammeter. Suitable switching arrangements are required for multiple-range ammeters. The Ayrton shunt ammeter is popular because special switches are not required to make basic dc current measurements. A very useful tool for simplifying circuits of moderate complexity is Thevenin's theorem. This theorem is often used to quickly determine if a particular measurement is accurate. Thevenin's theorem is reviewed in Appendix E for those readers requiring a more detailed discussion than provided in the example problems. The loading error will not be significant if $R_{TH} \gg R_m$. If R_{TH} is not known or easily determined, good accuracy is ensured if $R \gg R_m$, where R is the resistance in the circuit directly in series with the ammeter.

3-8 QUESTIONS

1. Why is there a practical limit on the smallest size for R_{sh}?

2. Specify an elementary design problem where the ability to solve a system of linear algebraic simultaneous equations would prove useful.

3. What determines how "good" an instrument is?

4. What is another name for the amount of inaccuracy?

5. What causes ammeter loading error? How can the effects of loading error be minimized?

6. Dr. Courtine has stated that loading error is *always* more significant than calibration error *when* up-scale readings are involved. Do you agree? (Hint: Look closely at Example 3-7).

3-9 PROBLEMS

1. A circuit has a Thevenin resistance of 19 kΩ. What is the largest value R_m can have so that the loading error does not exceed 5 percent?

2. A 1-mA, 100-Ω meter movement is available. Determine the shunt resistance required to measure currents up to 10 mA. Draw an equivalent circuit for the shunted meter movement.

3. Let $R = 22.2$ Ω and $V = 266.4$ mV in Fig. 3-14A. Determine the percent accuracy and loading error if the ammeter in Problem 2 is used to measure I. Assume that calibration error is negligible.

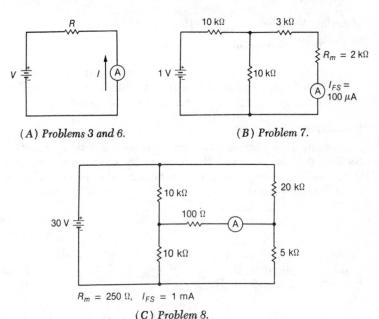

(A) Problems 3 and 6.

(B) Problem 7.

(C) Problem 8.

Fig. 3-14. Schematic diagrams for Chapter 3 problems.

4. Assume the accuracy of the meter movement in Problem 3 is 4 percent of the full-scale value. Within what range should the measured current in Problem 3 lie?

5. Assume that total error is the combined effect of calibration error and loading error. What is the worst case total error you would expect in Problem 4? Express your answer as a percentage.

6. What is the largest value R could have in Fig. 3-14A if the ammeter in Problem 1 is used to measure I and if the loading error is to be 1 percent or less?

7. Determine the percent accuracy and loading error in Fig. 3-14B. Assume that the calibration error is negligible.

8. What will the ammeter in Fig. 3-14C indicate? What is the current through the 100-Ω resistor if the ammeter is not in the circuit? In each case assume calibration error is negligible. (Hint: employ Thevenin's theorem.)

9. A 100-μA, 1-kΩ meter movement is available. Design an Ayrton shunt ammeter to provide the following ranges: 1 mA, 5 mA, and 50 mA.

10. Use the principle of current division to "check" your answers to Problem 9.

3-10   EXPERIMENT 3-1

Objective

The objective of this experiment is to construct a multiple-range dc ammeter.

Material Required

50-μA meter movement
9-V battery
Resistors or resistor decades. The values required will depend upon the values of I_{FS} and R_m for your meter movement. You will calculate the required values as part of the experiment.

Introduction

To measure a current (I_T) larger than I_{FS} a shunt resistance is connected in parallel with the basic meter movement. The value of shunt resistance required for a particular value of I_T can be calculated from Equation 3-1:

$$R_{sh} = \frac{I_{FS} R_m}{I_T - I_{FS}} \tag{3-1}$$

In this experiment you will determine the value of shunt resistance required to increase the range of your basic meter movement by factors of 5, 10, and 20. Thus

$$I_T = KI_{FS} \tag{3-9}$$

where K is called the *range factor*. Substituting Equation 3-9 into Equation 3-1 yields

$$R_{sh} = \frac{I_{FS} R_m}{KI_{FS} - I_{FS}}$$

$$R_{sh} = \frac{I_{FS} R_m}{I_{FS}(K-1)} = \frac{R_m}{K-1} \qquad (3\text{-}10)$$

Equation 3-10 enables you to *quickly* calculate the required values of shunt resistance. Thus, for range factors of 5, 10, and 20 the required values of shunt resistance are $R_m/4$, $R_m/9$, and $R_m/19$, respectively.

Procedure

Step 1. Calculate the shunt resistors required for range factors of 5, 10, and 20. Record your answers in Table 3-5.

Table 3-5. Required Shunt Resistors

Range Factor	I_T	R_{sh}
5	5(50 μA) = 250 μA	
10	10(50 μA) = 500 μA	
20	20(50 μA) = 1 mA	

Step 2. Select a convenient voltage source, V (a 9-V battery will do). Calculate the value of R in Fig. 3-15 so that $I_T = 250$ μA.

$$R = \frac{V}{I_T} = \underline{\hspace{3cm}}$$

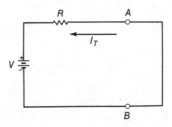

Fig. 3-15. Original circuit of Experiment 3-1.

Step 3. Build the circuit shown in Fig. 3-16. Record the value of I_m; ideally this should equal I_{FS}. *Caution:* Make sure the current through the meter movement is less than or equal to I_{FS}.

$I_m = \underline{\hspace{3cm}}$

Step 4. Repeat Steps 2 and 3 for range factors of 10 ($I_T = 500$ μA) and 20 ($I_T = 1$ mA).

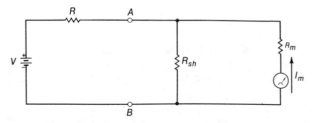

Fig. 3-16. Inserting the ammeter in Experiment 3-1.

Discussion

Some sources of error in this experiment include:

1. Calibration error
2. Loading error
3. Resistor tolerances
4. Battery voltage variations

If your battery provided a terminal voltage (under load) significantly different from 9 V, then a significant error was introduced into the experiment. You can compensate for such variations by measuring the (in-circuit) battery voltage in Fig. 3-15 and then adjusting R accordingly. For example, if the battery voltage in Fig. 3-15 measured 10 V, then R should be set to 10 V/I_T *prior* to taking data. An alternative approach, which ensures a stable voltage source, is illustrated in Fig. 3-17. In Fig. 3-17 we have employed a zener diode to provide a stable, well-regulated source voltage for the experiment. Resistor R_1 is chosen to limit the zener current to a safe value. The value of R_1 can be determined from

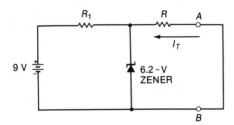

Fig. 3-17. A stable voltage source.

$$R_1(\max) = \frac{V_{in}(\min) - V_z}{I_L(\max)} \qquad (3\text{-}11)$$

Since you are employing a 9-V battery for V_{in} it is reasonable to assume $V_{in}(\min) \cong 8$ V. Current $I_L(\max)$ is approximately equal to I_T, and $V_z = 6.2$ V for the zener diode employed in Fig. 3-17. Thus

$$R_1(\text{max}) = \frac{8\text{ V} - 6.2\text{ V}}{I_L(\text{max})} = \frac{1.8\text{ V}}{I_T}$$

The value of R in Fig. 3-17 is 6.2 V/I_T. Remember that I_T is *different* for *each* range.

If your initial data indicates significant error due to variations in battery voltage, employ one of these methods and perform the experiment again.

Conclusion

How significant are loading error, calibration error, and resistor tolerances in this experiment? How can these errors be minimized?

3-11 EXPERIMENT 3-2

Objective

The objective of this experiment is to design an Ayrton shunt ammeter that provides the following ranges:

Range 1: $0-5I_{FS} = 0-250\ \mu\text{A}$
Range 2: $0-10I_{FS} = 0-500\ \mu\text{A}$
Range 3: $0-20I_{FS} = 0-1\ \text{mA}$

See Fig. 3-18.

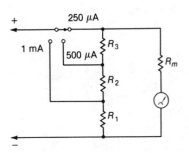

Fig. 3-18. Ayrton shunt ammeter for Experiment 3-2.

Material Required

50-μA meter movement
9-V battery
Resistor or resistor decades. The values required will depend on the values of I_{FS} and R_m for your meter movement. You will calculate the required values as part of the experiment.
The ability to solve a system of simultaneous linear algebraic equations.

Introduction

The ranges specified for the Ayrton shunt ammeter in this experiment are the same as those employed for the simple ammeter you designed in Experiment 3-1. Unlike the simple ammeter in Experiment 3-1, the Ayrton shunt ammeter is protected with a shunt resistance across the meter movement when you switch ranges. This makes the instrument easy to use and reduces the chance of damaging the meter movement due to an overload. The design of an Ayrton shunt ammeter is more tedious than a simple ammeter because you must solve three simultaneous equations to determine appropriate values for R_1, R_2, and R_3. Example 3-4 illustrated the design procedure for an Ayrton shunt ammeter. This approach can be modified slightly to include the concept of range factor introduced in Experiment 3-1. Since some users prefer this method, we will rework Example 3-4 utilizing the concept of range factor.

Determine the range factor for each range.

1-mA range:

$$K = \frac{1 \text{ mA}}{50 \text{ } \mu\text{A}} = 20$$

10-mA range:

$$K = \frac{10 \text{ mA}}{50 \text{ } \mu\text{A}} = 200$$

100-mA range:

$$K = \frac{100 \text{ mA}}{50 \text{ } \mu\text{A}} = 2\,000$$

Examine each switch position (Fig. 3-5).

Position 1 (1 mA):

$$R_{mE} = R_m = 2\text{k}\Omega$$

$$R_{shE} = R_1 + R_2 + R_3 = \frac{R_{mE}}{K-1} = \frac{2 \text{ k}\Omega}{19} = 105.3 \text{ } \Omega$$

Position 2 (10 mA):

$$R_{mE} = R_m + R_3 = 2 \text{ k}\Omega + R_3$$

$$R_{shE} = R_1 + R_2 = \frac{R_{mE}}{K-1} = \frac{2 \text{ k}\Omega + R_3}{199}$$

Since $R_1 + R_2 + R_3 = 105.3 \text{ } \Omega$,

$$R_1 + R_2 = 105.3 \text{ } \Omega - R_3$$

Thus

$$105.3 \text{ } \Omega - R_3 = \frac{2 \text{ k}\Omega + R_3}{199}$$

$$199(105.3\,\Omega - R_3) = 2\,k\Omega + R_3$$
$$20\,954.7\,\Omega - 199R_3 = 2\,k\Omega + R_3$$
$$200R_3 = 18\,954.7\,\Omega$$
$$R_3 = \frac{18\,954.7\,\Omega}{200} = 94.8\,\Omega$$

Position 3 (100 mA):

$$R_{mE} = R_m + R_2 + R_3 = 2\,k\Omega + R_2 + 94.8\,\Omega$$
$$= 2\,094.8\,\Omega + R_2$$
$$R_{shE} = R_1 = \frac{R_{mE}}{1\,999} = \frac{2\,094.8\,\Omega + R_2}{1\,999}$$

Since $R_1 + R_2 + R_3 = 105.3\,\Omega$

$$R_1 = 105.3\,\Omega - R_2 - R_3$$
$$= 105.3\,\Omega - R_2 - 94.8\,\Omega$$
$$= 10.5\,\Omega - R_2$$

Thus

$$10.5\,\Omega - R_2 = \frac{2\,094.8\,\Omega + R_2}{1\,999}$$
$$1\,999(10.5\,\Omega - R_2) = 2\,094.8\,\Omega + R_2$$
$$20\,989.5\,\Omega - 1\,999R_2 = 2\,094.8\,\Omega + R_2$$
$$2\,000R_2 = 18\,894.7\,\Omega$$
$$R_2 = \frac{18\,894.7\,\Omega}{2\,000} = 9.45\,\Omega$$

Finally,

$$R_1 + R_2 + R_3 = 105.3\,\Omega$$
$$R_1 + 9.45\,\Omega + 94.8\,\Omega = 105.3\,\Omega$$
$$R_1 = 1.05\,\Omega$$

Procedure

Step 1. Utilize the procedure of Example 3-4, or the modified procedure provided in this experiment, to design the ammeter.

Step 2. Before building the ammeter, *verify your design "on paper"* by employing the current division principle to see if full-scale deflection of the basic meter movement occurs for each switch position.

Step 3. Determine appropriate values for V and R in Fig. 3-19 to produce currents of 250 μA, 500 μA, and 1 mA. (The circuit of Experiment 3-1 in Fig. 3-17 can be used here.)

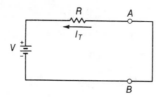

Fig. 3-19. Test circuit for Experiment 3-2.

Range 1: $V =$ _____ $R =$ _____

Range 2: $V =$ _____ $R =$ _____

Range 3: $V =$ _____ $R =$ _____

Step 4. Build the circuit shown in Fig. 3-20. Record the value of I_m for each range. Remember, R must be changed when you switch ranges. Be *careful* so that you do not damage the meter movement.

Range 1: $I_m =$ _____

Range 2: $I_m =$ _____

Range 3: $I_m =$ _____

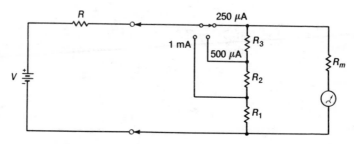

Fig. 3-20. Inserting the ammeter in Experiment 3-2.

Discussion

The ultimate test of any design is simple—does it work? If you solved the simultaneous equations correctly and employed component values close to the calculated values, then the ammeter should have worked properly. The ability to solve elementary equations is useful in design and often in providing an insight into how a circuit works. The reader is encouraged to learn the basic math skills necessary to understand electronic instruments and systems.

Conclusion

Compare the ammeters designed in Experiments 3-1 and 3-2. Could Ayrton shunt ammeters be designed to provide more than three ranges? How would the design procedure for such an ammeter compare with the one you employed in this experiment?

3-12 EXPERIMENT 3-3

Objective

The objective of this experiment is to investigate ammeter loading effects.

Material Required

The Ayrton shunt ammeter you designed in Experiment 3-2
9-V battery
150-Ω resistor
1-k Ω potentiometer
Oscilloscope or sensitive dc voltmeter

Introduction

Ammeter accuracy (a) and loading error (e) are defined by Equations 3-5 and 3-7, respectively. Specifically,

$$a = \frac{R_{TH}}{R_{TH} + R_m} \qquad (3\text{-}5)$$

$$e = 1 - a \qquad (3\text{-}7)$$

Equations 3-5 and 3-7 deal only with *loading* error. To determine total error you have to take into account *all* sources of error. Since loading error often constitutes most of the total error, Equations 3-5 and 3-7 serve as approximations for total error. Note that ammeter loading error is minimized when $R_{TH} \gg R_m$.

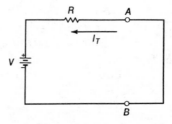

Fig. 3-21. Original circuit of Experiment 3-3.

Procedure

Step 1. Assume V in Fig. 3-21 to be 5 V. Determine the value of R required to produce an I_T of 250 μA, 500 μA, and 1 mA. Repeat for $V = 0.5$ V. Record answers on the data sheet at the end of this experiment (Fig. 3-23).

Step 2. Using the value of R calculated for $V = 5$ V, $I_T = 250$ μA, build the circuit shown in Fig. 3-22. With terminals AB

73

shorted, adjust the 1-kΩ potentiometer until $V = 5$ V. An oscilloscope is recommended for this and similar adjustments.

Step 3. Remove the short between *AB*. Measure the current in the branch *AB* using your Ayrton shunt ammeter from Experiment 3-2. Remember to use the *appropriate range* for the measurement.

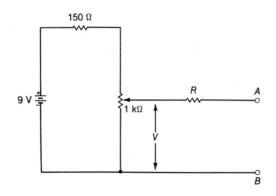

Fig. 3-22. Test circuit in Experiment 3-3.

Step 4. Remove the ammeter and substitute the value of *R* calculated for $V = 0.5$ V, $I_T = 250$ μA. With terminals *AB* shorted, adjust the 1-kΩ potentiometer until $V = 0.5$ V.

Step 5. Remove the short between *AB*. Measure the current in the branch *AB* with your Ayrton shunt ammeter.

Step 6. Follow a similar procedure for the remaining two ranges, 0–500 μA and 0–1 mA. Make sure that $I_m \leqq I_{FS}$.

Step 7. From your data calculate the *measured* accuracy [(I_m/I_{FS}) × 100%] for each measurement.

Step 8. Compare your measured accuracies with those obtained from Equation 3-5. In Equation 3-5 $R_{TH} \cong R$ and $R_m = R_{shE} \parallel R_{mE}$.

Conclusion

Why were some measurements accurate and other measurements inaccurate? What are significant sources of error in this experiment? Are the sources of error you identified sufficient to explain the observed differences between measured and calculated values?

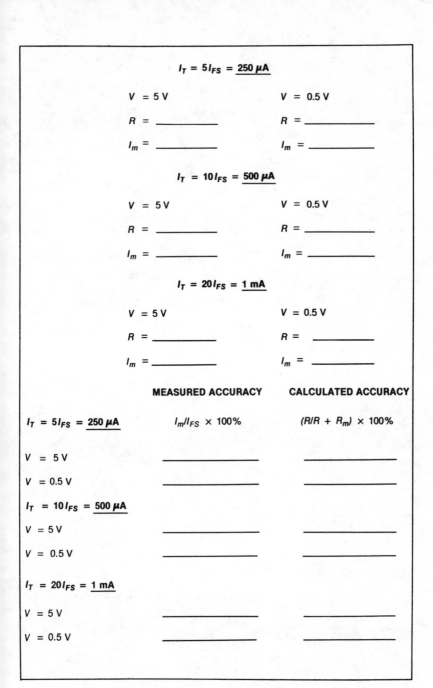

$I_T = 5I_{FS} = \underline{250\ \mu A}$

$V = 5\ V$	$V = 0.5\ V$
$R =$ _____	$R =$ _____
$I_m =$ _____	$I_m =$ _____

$I_T = 10I_{FS} = \underline{500\ \mu A}$

$V = 5\ V$	$V = 0.5\ V$
$R =$ _____	$R =$ _____
$I_m =$ _____	$I_m =$ _____

$I_T = 20I_{FS} = \underline{1\ mA}$

$V = 5\ V$	$V = 0.5\ V$
$R =$ _____	$R =$ _____
$I_m =$ _____	$I_m =$ _____

	MEASURED ACCURACY	CALCULATED ACCURACY
$I_T = 5I_{FS} = \underline{250\ \mu A}$	$I_m/I_{FS} \times 100\%$	$(R/R + R_m) \times 100\%$
$V = 5\ V$	_____	_____
$V = 0.5\ V$	_____	_____
$I_T = 10I_{FS} = \underline{500\ \mu A}$		
$V = 5\ V$	_____	_____
$V = 0.5\ V$	_____	_____
$I_T = 20I_{FS} = \underline{1\ mA}$		
$V = 5\ V$	_____	_____
$V = 0.5\ V$	_____	_____

Fig. 3-23. Data sheet for Experiment 3-3.

CHAPTER 4

DC Voltmeters

4-1 INTRODUCTION

At this point you should have a good understanding of how dc ammeters work, how they are designed, and the factors that determine how accurate (or inaccurate) a particular measurement will be. You are now ready to consider the problem of measuring dc voltages. Generally speaking, voltages are easier to measure than currents. This is because voltmeters are connected in *parallel* with the load, rather than in series with the load. Thus, when you measure voltage the circuit does not have to be "broken" to insert the voltmeter. Perhaps this fact alone makes the voltmeter the most popular troubleshooting instrument.

4-2 OBJECTIVES

At the end of this chapter you will be able to do the following:

- Determine the series resistance (R_s) required to convert an ammeter into a voltmeter.
- Define voltmeter sensitivity.
- Design simple multirange dc voltmeters with a series arrangement of multiplier resistors.
- Estimate the percent accuracy and loading error for a particular measurement.
- Specify the conditions necessary to minimize voltmeter loading error.

4-3 AN ELEMENTARY DC VOLTMETER

A basic meter movement can be converted to a dc voltmeter with the addition of a *series* resistance (R_s). As you will see, this series

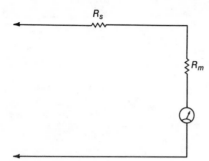

Fig. 4-1. An elementary dc voltmeter.

resistance is often called a *multiplier*. The basic instrument is illustrated in Fig. 4-1. Recall that the meter movement provides full-scale deflection when the current through it equals I_{FS}. You wish to measure voltages up to some maximum, which is the full-scale value (V_{FS}). Thus, the idea is to pick a value for R_s so that when V_{FS} is applied to the input of the voltmeter the resulting current equals I_{FS}. Of course, the meter face is calibrated to read voltage rather than current. This concept will be illustrated shortly via a pedagogically sound example.

4-4 DESIGN BASICS

The elementary dc voltmeter is illustrated in Fig. 4-2, where you wish to determine the value of R_s. Referring to Fig. 4-2 we will derive several useful relationships and then illustrate their application to the design of dc voltmeters.

The equivalent *input* resistance of the voltmeter (R_{in}) by inspection is

$$R_{in} = R_s + R_m \qquad (4\text{-}1)$$

From Ohm's law,

$$R_{in} = \frac{V_{FS}}{I_{FS}} \qquad (4\text{-}2)$$

Therefore

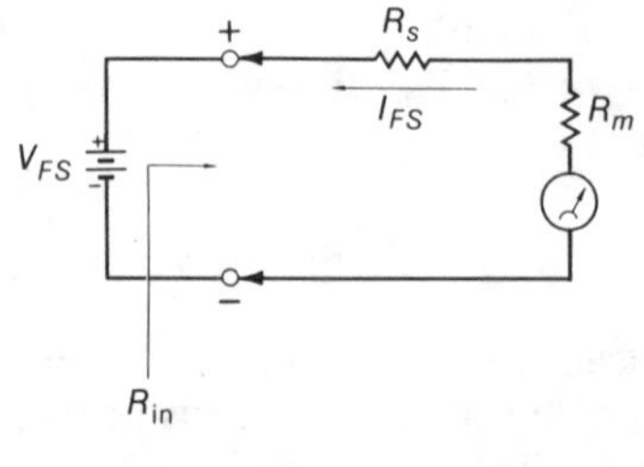

Fig. 4-2. Calculating R_s.

$$R_s + R_m = \frac{V_{FS}}{I_{FS}}$$

or

$$R_s = \frac{V_{FS}}{I_{FS}} - R_m \qquad (4\text{-}3)$$

Recall that the amount of current required to produce full-scale deflection (I_{FS}) is a good indicator of the sensitivity of a meter movement. The dc sensitivity of a voltmeter is given in terms of "ohms per volt." Thus, *dc voltmeter sensitivity* (S) is defined as the reciprocal of I_{FS}:

$$S = \frac{1}{I_{FS}} \qquad (4\text{-}4)$$

where S is given in ohms per volt (Ω/V). Substituting Equation 4-4 into Equations 4-2 and 4-3 yields

$$R_{in} = SV_{FS} \qquad (4\text{-}5)$$

$$R_s = SV_{FS} - R_m \qquad (4\text{-}6)$$

Equations 4-5 and 4-6 are sufficient to design an elementary dc voltmeter. Now for the pedagogically sound example alluded to previously.

EXAMPLE 4-1

You have just received a 50-μA, 2-kΩ basic meter movement as a gift. Design a 0- to 10-V dc voltmeter. Include the following information:
(a) The sensitivity of the voltmeter.
(b) The input resistance of the voltmeter.
(c) The value of the series resistor.
(d) A sketch illustrating how the meter face was marked when you received it.
(e) A sketch illustrating how the meter face can be calibrated to read voltage.

(a)

$$S = \frac{1}{I_{FS}} = \frac{1}{50 \ \mu\text{A}} = 20 \ \frac{\text{k}\Omega}{\text{V}}$$

(b)

$$R_{in} = SV_{FS}$$
$$= \left(20 \frac{\text{k}\Omega}{\text{V}} \right) (10 \text{ V}) = 200 \text{ k}\Omega$$

(c)

$$R_s = R_{in} - R_m$$
$$= 200 \text{ k}\Omega - 2 \text{ k}\Omega$$
$$= 198 \text{ k}\Omega$$

(d) The meter face of your meter movement is illustrated in Fig. 4-3.

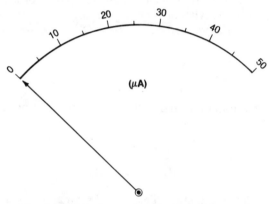

Fig. 4-3. Original meter face for Example 4-1.

(e) The amount of deflection (D) depends on the current through the meter movement (I_m). Thus you can define deflection D as

$$D = \frac{I_m}{I_{FS}} \qquad (4\text{-}7)$$

In order to calibrate the meter face to read voltage you need to know the relationship between input voltage (V_{in}) and the amount of deflection (D) produced by applying V_{in} to the voltmeter. Fortunately this relationship is easily arrived at:

$$V_{in} = I_m R_{in}$$

Substituting $D I_{FS}$ for I_m yields

$$V_{in} = D I_{FS} R_{in} \qquad (4\text{-}8)$$

Equation 4-7 enables us to construct Table 4-1, which indicates the amount of current required to produce deflections of 0, 20, 40, 60, 80, and 100 percent. Similarly, Equation 4-8 is employed to generate Table 4-2, which indicates the relationship between input voltage (V_{in}) and the percent deflection ($D \times 100$). The calibrated voltage scale is illustrated in Fig. 4-4, and the voltmeter in Fig. 4-5.

EXAMPLE 4-2

Determine the input voltage required for full-scale deflection if $R_s = 0$ for the voltmeter in Example 4-1.

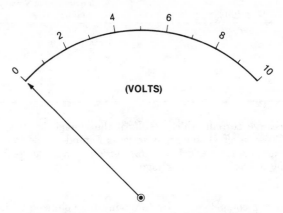

Fig. 4-4. Calibrated voltage scale for Example 4-1.

Fig. 4-5. Dc voltmeter for Example 4-1.

Table 4-1. Meter Current (I_m) for Various Deflections in Example 4-1

Percent Deflection $D \times 100\%$	Meter Current $I_m = DI_{FS}$
100	50 μA
80	40 μA
60	30 μA
40	20 μA
20	10 μA
0	0 μA

$$V_{in} = DI_{FS}R_{in}$$
$$= (1)(50\ \mu A)(2\ k\Omega)$$
$$= 100\ mV$$

This example illustrates the fact that a basic meter movement with *no* external series resistance can be used to measure "small" voltages. Frequently a separate jack is provided for this range, as well as for a high-voltage range. This is done to protect the meter movement

81

Table 4-2. Input Voltage (V_{in}) for Various Deflections in Example 4-1

Percent Deflection $D \times 100\%$	Required Input Voltage $V_{in} = DI_{FS}R_{in}$
100	10 V
80	8 V
60	6 V
40	4 V
20	2 V
0	0 V

from excessive current which results if the voltage being measured exceeds V_{FS}. In effect, the separate jacks remind the person making the measurement to *select the appropriate range carefully* when measuring a small or large voltage.

EXAMPLE 4-3

You have a compelling desire to modify the voltmeter of Example 4-1 in order to make it more useful. To meet this need design the voltmeter to provide the following ranges: 0–0.1, 0–10, 0–50, 0–250, and 0–1000 V.

The "design procedure" is straightforward:

1. *Calculate the sensitivity:*

$$S = \frac{1}{I_{FS}} = \frac{1}{50 \ \mu A} = 20 \ \frac{k\Omega}{V}$$

2. *Calculate R_{in} for each range:*

$$R_{in} = SV_{FS}$$

3. *Calculate R_s for each range:*

$$R_s = R_{in} - R_m$$

The results of this three-step procedure are summarized in Table 4-3. The addition of a rotary switch completes the design, which is illustrated in Fig. 4-6. Note in Fig. 4-6 that separate jacks have been provided for the low (0–0.1 V) and high (0–1000 V) ranges.

Table 4-3. Values of R_{in} and R_s for Each Range in Example 4-3

Range (V_{FS})	R_{in} (SV_{FS})	R_s ($R_{in} - R_m$)
0–0.1 V	2 kΩ	0 Ω
0–10 V	200 kΩ	198 kΩ
0–50 V	1 MΩ	0.998 MΩ
0–250 V	5 MΩ	4.998 MΩ
0–1000 V	20 MΩ	19.998 MΩ

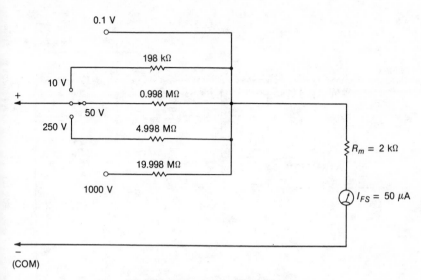

Fig. 4-6. Voltmeter for Example 4-3.

4-5 SERIES MULTIPLIER DESIGN

An arrangement of multiplier resistors similar to the one shown in Fig. 4-6 is not usually found in commercial instruments. A *series arrangement* of multiplier resistors is used instead. An alternate design for Example 4-3 utilizing a series arrangement of multiplier resistors is illustrated in Fig. 4-7. Several additional features are also included in Fig. 4-7 and will be discussed shortly. Table 4-4 indicates which resistors in Fig. 4-7 are switched in series with the meter movement for each range. The values of R_s in Table 4-4 were calculated in Example 4-3. The values of the series multiplier resistors are easily determined from Table 4-4 as follows:

$$R_4 = 198 \text{ k}\Omega$$

and

$$R_3 + R_4 = 998 \text{ k}\Omega$$

Therefore

$$\begin{aligned} R_3 &= 998 \text{ k}\Omega - R_4 \\ &= 998 \text{ k}\Omega - 198 \text{ k}\Omega \\ &= 800 \text{ k}\Omega \end{aligned}$$

Next,

$$R_2 + R_3 + R_4 = 4.998 \text{ M}\Omega$$

83

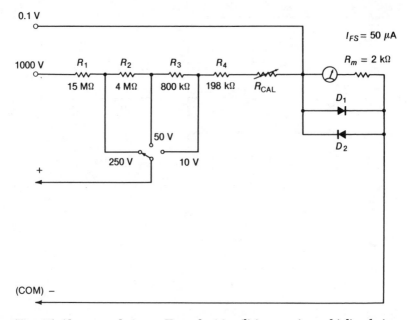

Fig. 4-7. Alternate solution to Example 4-3 utilizing a series multiplier design.

$$R_2 = 4.998 \text{ M}\Omega - R_3 - R_4$$
$$= 4.998 \text{ M}\Omega - 0.8 \text{ M}\Omega - 0.198 \text{ M}\Omega$$
$$= 4 \text{ M}\Omega$$

Finally,

$$R_1 + R_2 + R_3 + R_4 = 19.998 \text{ M}\Omega$$
$$R_1 = 19.998 \text{ M}\Omega - R_2 - R_3 - R_4$$
$$= 19.998 \text{ M}\Omega - 4 \text{ M}\Omega - 0.8 \text{ M}\Omega - 0.198$$
$$= 15 \text{ M}\Omega$$

The reason for the popularity of the series arrangement of multiplier resistors is due to the fact that all resistors (multipliers) except

Table 4-4. Values of R_s for Each Range in Fig. 4-7

Range	R_s	Value of R_s
0–0.1 V	0	0 Ω
0–10 V	R_4	198 kΩ
0–50 V	$R_3 + R_4$	0.998 MΩ
0–250 V	$R_2 + R_3 + R_4$	4.998 MΩ
0–1 000 V	$R_1 + R_2 + R_3 + R_4$	19.998 MΩ

R_4 can usually be matched to *standard* resistance values. This is a significant advantage over the arrangement used in Fig. 4-6 which often requires "oddball" resistance values for *each* multiplier. The discriminating reader and/or figure watcher has undoubtedly noticed the addition of diodes D_1 and D_2, as well as the resistor labeled R_{CAL} in Fig. 4-7. Notice that both diodes are in parallel with the meter movement. Thus the voltage across each diode is the same as the voltage across the meter movement ($I_m R_m$). When full-scale deflection occurs the voltage across the diodes is $(50 \ \mu A) \times (2 \ k\Omega)$, or 0.1 V. Thus diodes D_1 and D_2 are normally reversed biased and act like open circuits. In the event of an overload ($V_{in} > V_{FS}$), however, the current through the meter movement will exceed 50 μA. This condition produces a voltage across the meter movement (and diodes) *greater* than 0.1 V. Depending on the polarity of the input voltage, the amount of overload, and the knee voltage of the diodes either D_1 or D_2 will become forward biased, thus bypassing most of the current around the meter movement. Diodes D_1 and D_2 should have relatively small knee voltages. In any case you should avoid overloads! Even with the protective diodes in the circuit an overload could still damage the meter movement by bending the pointer as it attempts to move past the stops (retaining pins) on either side of the meter face.

As the name implies, the resistor labeled R_{CAL} in Fig. 4-7 is used to calibrate the voltmeter. The internal resistances of meter movements (of the same type) vary slightly from one unit to the next. In addition, component values change with temperature, aging, etc. Resistor R_{CAL} is used to compensate for such variations.

4-6 VOLTMETER LOADING EFFECTS

In Chapter 3 we discussed ammeter loading effects. A similar analysis for dc voltmeters is highly desirable. That is of course the purpose of this section.

To begin, consider the voltage across R_2 in Fig. 4-8A. Without a voltmeter in the circuit the *original* voltage across R_2 is the Thevenin voltage, V_{TH}. This startling fact is indicated in Fig. 4-8B. When you employ a voltmeter to measure the voltage across R_2, the input resistance ($R_{in} = R_m + R_s$) of the voltmeter is placed in parallel with R_2, as indicated in Figs. 4-9A and 4-9B. Thus the voltage across R_2 is *changed* by the presence of the voltmeter. Fig. 4-10 represents the Thevenin equivalent circuit (Fig. 4-8B) *with* the voltmeter (load) connected. In order to keep our results general we will discuss the Thevenin equivalent circuits in Figs. 4-8B and 4-10.

The *original* voltage (V_o) from Fig. 4-8B is $V_o = V_{TH}$. Similarly, the *measured* voltage (V_m) from Fig. 4-10 is

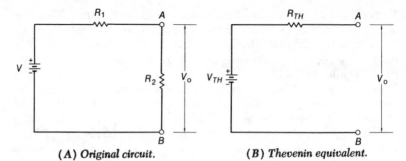

(A) Original circuit.   (B) Thevenin equivalent.

Fig. 4-8. Original and Thevenin equivalent circuits discussed in text.

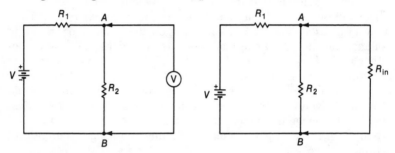

(A) Voltmeter inserted.   (B) Effect of R_{in}.

Fig. 4-9. Employing a voltmeter to measure the voltage across R_2.

Fig. 4-10. Thevenin equivalent circuit with voltmeter (load) connected.

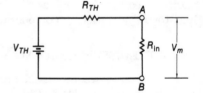

$$V_m = V_{TH}\left(\frac{R_{in}}{R_{TH} + R_{in}}\right)$$

Hence the *accuracy* associated with the measurement (*a*) is

$$a = \frac{V_m}{V_o} = V_{TH}\left(\frac{R_{in}}{R_{TH} + R_{in}}\right) \div V_{TH} = \frac{R_{in}}{R_{in} + R_{TH}} \qquad (4\text{-}9)$$

The *error* (*e*) is given by

$$e = \frac{V_o - V_m}{V_o} \qquad (4\text{-}10)$$

Remember that you can think of *error* as the amount of inaccuracy. Thus

$$e = 1 - a \qquad (4\text{-}11)$$

The above relationships are of course analogous to those obtained in our discussion of ammeter loading effects.

Once again you will consider *good accuracy* to prevail if the resulting error is $e \leqq 10$ percent. In order to determine the relative size of R_{in} for good accuracy we solve Equation 4-9 for R_{in} as follows:

$$a = \frac{R_{in}}{R_{in} + R_{TH}}$$

$$aR_{in} + aR_{TH} = R_{in}$$

$$R_{in} - aR_{in} = aR_{TH}$$

$$R_{in}(1 - a) = aR_{TH}$$

$$R_{in} = \frac{aR_{TH}}{1 - a}$$

Substituting e for $1 - a$ yields

$$R_{in} = \frac{aR_{TH}}{e} \qquad (4\text{-}12)$$

Equation 4-12 is employed to generate Table 4-5, which indicates the relative size of R_{in} for good accuracy. Even a cursory examination of Table 4-5 indicates that good accuracy results when $R_{in} \gg R_{TH}$.

Table 4-5. Relative Size of R_{in} for Good Voltmeter Accuracy

Percent Accuracy ($a \times 100$)	Percent Error ($e \times 100$)	$R_{in} = \dfrac{aR_{TH}}{e}$
99	1	$99R_{TH}$
98	2	$49R_{TH}$
97	3	$32.3R_{TH}$
96	4	$24R_{TH}$
95	5	$19R_{TH}$
94	6	$15.7R_{TH}$
93	7	$13.3R_{TH}$
92	8	$11.5R_{TH}$
91	9	$10.1R_{TH}$
90	10	$9R_{TH}$

Example 4-4

The voltmeter of Example 4-3 is used to measure the voltage between A and B in Fig. 4-11. For the following calculations (a) through (d) assume that the loading error is the only significant source of error. Determine:

(a) The Thevenin equivalent circuit.
(b) The original voltage.

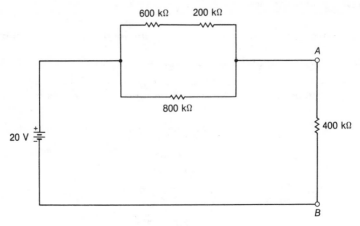

Fig. 4-11. Circuit for Example 4-4.

(c) The measured voltage assuming the 0- to 10-V range is used for the measurement.

(d) The measured voltage assuming the 0- to 50-V range is used for the measurement.

(e) Predict the ranges within which the measured values should lie, assuming the meter movement has an accuracy of 3 percent of the full-scale value.

(f) Discuss the results obtained in (a) through (e).

Solutions:

(a) First you determine R_{TH}. The details of this process are shown in Fig. 4-12.

$$R_{TH} = (600 \text{ k}\Omega + 200 \text{ k}\Omega) \ || \ 800 \text{ k}\Omega \ || \ 400 \text{ k}\Omega$$
$$= 200 \text{ k}\Omega$$

Next you determine V_{TH}. From Fig. 4-13, V_{TH} is determined to be 10 V. The Thevenin equivalent circuit is illustrated in Fig. 4-14.

(b) The original voltage between A and B equals V_{TH}. Thus $V_o = 10$ V.

(c) When the voltmeter is placed in the circuit R_{in} is connected in parallel with the branch AB. For the 0- to 10-V range

$$R_{in} = SV_{FS}$$
$$= \left(20 \frac{\text{k}\Omega}{\text{V}}\right)(10 \text{ V}) = 200 \text{ k}\Omega$$

This is illustrated in Fig. 4-15A. Therefore

$$V_m = \frac{(10 \text{ V})(200 \text{ k}\Omega)}{200 \text{ k}\Omega + 200 \text{ k}\Omega} = 5 \text{ V}$$

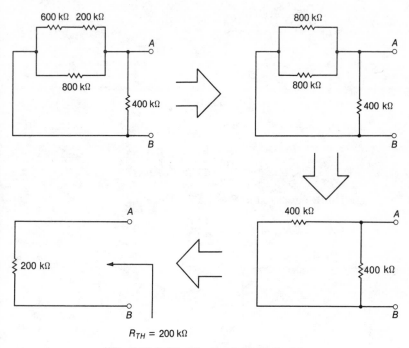

Fig. 4-12. Solving for R_{TH} in Example 4-4.

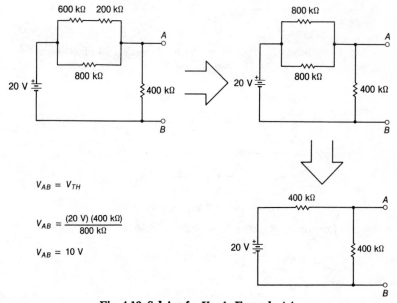

$V_{AB} = V_{TH}$

$V_{AB} = \dfrac{(20\ V)\ (400\ k\Omega)}{800\ k\Omega}$

$V_{AB} = 10\ V$

Fig. 4-13. Solving for V_{TH} in Example 4-4.

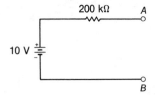

Fig. 4-14. Thevenin equivalent circuit
for Example 4-4.

(*d*) For the 0- to 50-V range

$$R_{in} = SV_{FS}$$

$$= \left(20 \frac{k\Omega}{V}\right)(50 \text{ V}) = 1 \text{ M}\Omega$$

This is illustrated in Fig. 4-15B. Therefore

$$V_m = \frac{(10 \text{ V})(1 \text{ M}\Omega)}{200 \text{ k}\Omega + 1 \text{ M}\Omega} = 8.33 \text{ V}$$

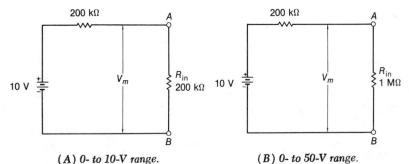

(A) 0- to 10-V range.   (B) 0- to 50-V range.

Fig. 4-15. Thevenin equivalent circuits for Example 4-4 with voltmeters
(loads) inserted.

(*e*) For the 0- to 10-V range, 3 percent of 10 V is

$$\frac{3}{100} \times 10 \text{ V} = 0.3 \text{ V}$$

The expected range R is

$$R = V_m \pm CE$$
$$= 5 \text{ V} \pm 0.3 \text{ V}$$
$$= 4.7 \text{ V to } 5.3 \text{ V}$$

For the 50-V range, 3 percent of 50 V is

$$\frac{3}{100} \times 50 \text{ V} = 1.5 \text{ V}$$

Therefore

$$R = V_m \pm CE$$

$$= 8.33 \text{ V} \pm 1.5 \text{ V}$$
$$= 6.83 \text{ V to } 9.83 \text{ V}$$

(f) Discussion: The voltage between A and B in the original circuit is 10 V. When the voltmeter is placed in the circuit its internal resistance is placed in parallel with the resistance between the points of measurement. The resistance of the circuit between these two points is lowered. Thus there is a smaller voltage drop between these points. In this example, if the 0- to 10-V range is used for the measurement, the circuit is changed in such a manner that V_{AB} *becomes* 5 V. Similarly, V_{AB} becomes 8.33 V when the 0- to 50-V range is employed for the measurement. Neglecting calibration and any other sources of error, 5 V and 8.33 V are the *actual* voltages between A and B *when* the voltmeter is placed in the circuit employing the 0- to 10-V and 0- to 50-V ranges, respectively. The voltmeter would *read* 5 V on the 0- to 10-V range and 8.33 V on the 0- to 50-V range *if* loading error were the *only* source of error.

By considering the combined effects of calibration and loading error you can predict quite accurately the range within which your measured values should lie. On either the 0- to 10-V or 0- to 50-V range the measured values in this example problem would be *inaccurate*. This does *not* mean the voltmeter employed for the measurement was defective or poorly designed. Poor accuracy resulted because R_{in} was *not* significantly larger than R_{TH}.

When you switch from the 0- to 10-V range to the 0- to 50-V range the input resistance (R_{in}) of the voltmeter increases from 200 kΩ to 1 MΩ. Thus there is *less* loading error on the 0- to 50-V range. You would observe a *significant change* in the voltmeter reading when you switched ranges! When such a change occurs, it is a clear indication that loading error *is* significant. *This point should be kept in mind when you troubleshoot circuits.*

Note that in this example, the effect of calibration error is more pronounced on the 0- to 50-V range than on the 0- to 10-V range. The difference between the maximum (5.3 V) and minimum (4.7 V) expected readings on the 0- to 10-V range when calibration error is considered is 0.6 V. Similarly, the difference between the maximum (9.83 V) and minimum (6.83 V) expected readings on the 0- to 50-V range is 3 V! Remember the significance of calibration error increases for down-scale readings. To ensure accurate readings you should observe the following guidelines.

1. When measuring a voltage (or current) select a range that results in deflection as close to full-scale as possible. This will *minimize* the effect of calibration error.

2. When measuring voltage select a range so that $R_{in} \gg R_{TH}$. This will minimize the effect of loading error. If

R_{TH} is not known or easily determined, then select a range so that $R_{in} \gg R$, where R is the resistance across which the voltage is being measured.

3. As a corollary to (2) if you are measuring a current, select a range so that $R_{TH} \gg R_m$ to minimize the effect of (ammeter) loading error. If R_{TH} is not known, then select a range where $R \gg R_m$, where R represents the resistance across which the voltage is being measured.

Example 4-5

Two voltmeters are employed to measure the voltage V_{AB} in Fig. 4-16A. The first voltmeter has a 500-μA meter movement, while

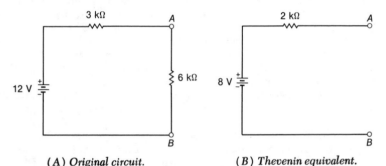

(A) Original circuit.　　　(B) Thevenin equivalent.

Fig. 4-16. Original and Thevenin equivalent circuits for Example 4-5.

the second voltmeter employs a 50-μA meter movement. Assume that the 0- to 10-V ranges are used for the measurements and that calibration error is negligible. Determine:

(a) What each voltmeter reads.

(b) The percent accuracy for each measurement.

(a) First you determine the Thevenin equivalent circuit. To obtain R_{TH} the 12-V source in Fig. 4-16A is shorted. Looking in between terminals A and B the equivalent resistance $R_{AB} = 6\text{ k}\Omega \parallel 3\text{ k}\Omega = 2\text{ k}\Omega$. Thus $R_{TH} = 2\text{ k}\Omega$. The voltage V_{TH} is simply the open-circuit voltage (remember, the voltmeters are considered loads). Thus $V_{TH} = V_{AB}$. Therefore

$$V_{TH} = (12\text{ V}) \left(\frac{6\text{ k}\Omega}{3\text{ k}\Omega + 6\text{ k}\Omega} \right) = 8\text{ V}$$

The Thevenin equivalent circuit is shown in Fig. 4-16B.

Next you calculate R_{in} for each voltmeter. For the first voltmeter (500-μA movement)

$$S_1 = \frac{1}{I_{FS}} = \frac{1}{500\ \mu\text{A}} = 2\ \frac{\text{k}\Omega}{\text{V}}$$

$$R_{in1} = S_1 V_{FS} = \left(2\ \frac{\text{k}\Omega}{\text{V}} \right) (10\text{ V}) = 20\text{ k}\Omega$$

Similarly, for the second voltmeter (50-μA movement)

$$S_2 = \frac{1}{I_{FS}} = \frac{1}{50 \ \mu A} = 20 \frac{k\Omega}{V}$$

$$R_{in2} = S_2 V_{FS} = \left(20 \frac{k\Omega}{V}\right)(10 \ V) = 200 \ k\Omega$$

When the voltmeters are placed in the circuit their R_{in}'s are placed across the branch AB. This is illustrated in Figs. 4-17A and 4-17B for each voltmeter. In Fig. 4-17A the 500-μA meter movement is used.

$$V_{m1} = (8 \ V) \left(\frac{20 \ k\Omega}{2 \ k\Omega + 20 \ k\Omega}\right) = 7.27 \ V$$

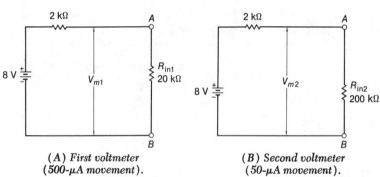

(A) First voltmeter
(500-μA movement).

(B) Second voltmeter
(50-μA movement).

Fig. 4-17. Thevenin equivalent circuit for Example 4-5 with voltmeters (loads) connected.

Similarly, in Fig. 4-17B, for the 50-μA meter movement

$$V_{m2} = (8 \ V) \left(\frac{200 \ k\Omega}{2 \ k\Omega + 200 \ k\Omega}\right) = 7.92 \ V$$

(b) The accuracy of the measurement employing the 500-μA meter movement is

$$a_1 = \frac{V_{m1}}{V_o} = \frac{V_{m1}}{V_{TH}} = \frac{7.27 \ V}{8 \ V} = 0.909 = 90.9\%$$

Similarly, when the 50-μA meter movement is used for the measurement,

$$a_2 = \frac{V_{m2}}{V_o} = \frac{V_{m2}}{V_{TH}} = \frac{7.92 \ V}{8 \ V} = 0.99 = 99\%$$

EXAMPLE 4-6

The voltmeters of Example 4-5 are used to measure V_{CD} in Fig. 4-18A. Determine (a) what each voltmeter reads and (b) the percent accuracy.

(a) The Thevenin equivalent circuit of Fig. 4-18A is shown in Fig. 4-18B. This was obtained in a manner similar to Ex-

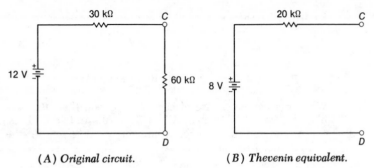

(A) Original circuit.　　　　(B) Thevenin equivalent.

Fig. 4-18. Original and Thevenin equivalent circuits for Example 4-6.

ample 4-5. Figs. 4-19A and 4-19B illustrate the Thevenin equivalent circuit with the R_{in} of each voltmeter connected across points C and D in the circuit. Thus, from Fig. 4-19A

$$V_{m1} = (8 \text{ V})\left(\frac{20 \text{ k}\Omega}{20 \text{ k}\Omega + 20 \text{ k}\Omega}\right) = 4 \text{ V}$$

From Fig. 4-19B

$$V_{m2} = (8 \text{ V})\left(\frac{200 \text{ k}\Omega}{20 \text{ k}\Omega + 200 \text{ k}\Omega}\right) = 7.27 \text{ V}$$

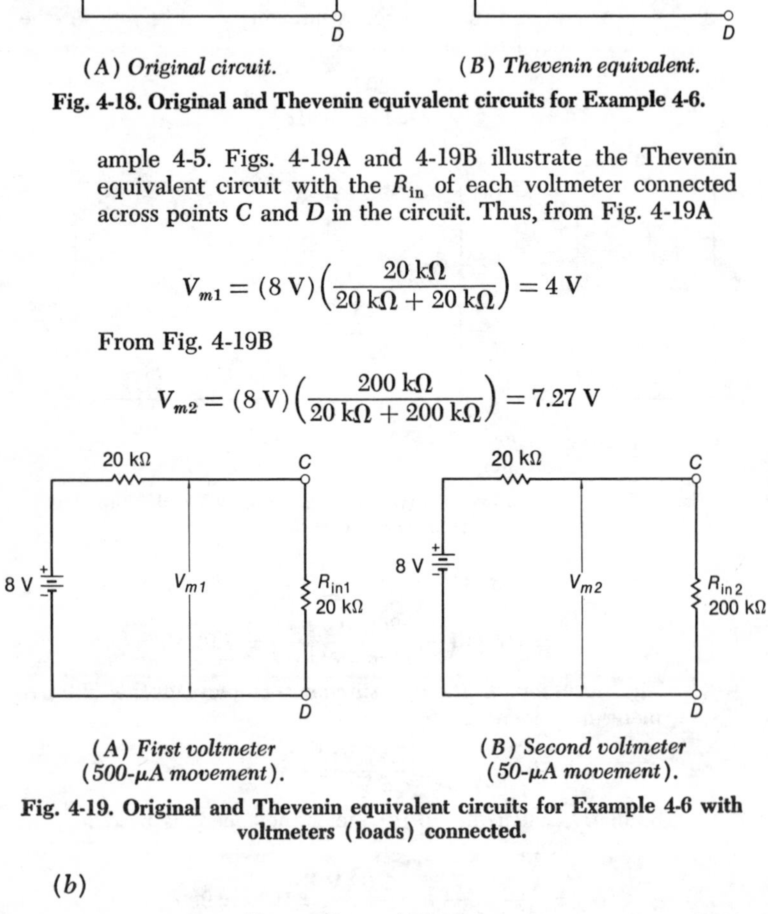

(A) First voltmeter　　　　(B) Second voltmeter
(500-μA movement).　　　　(50-μA movement).

Fig. 4-19. Original and Thevenin equivalent circuits for Example 4-6 with voltmeters (loads) connected.

(b)

$$a_1 = \frac{V_{m1}}{V_o} = \frac{V_{m1}}{V_{TH}} = \frac{4 \text{ V}}{8 \text{ V}} = 0.5 = 50\%$$

$$a_2 = \frac{V_{m2}}{V_o} = \frac{V_{m2}}{V_{TH}} = \frac{7.27 \text{ V}}{8 \text{ V}} = 0.909 = 90.9\%$$

Comparing Examples 4-5 and 4-6 you should note that both voltmeters provided accurate readings for the circuit in Fig. 4-16A. This was the case because the requirement for good voltmeter accuracy ($R_{in} \gg R_{TH}$) was satisfied for each voltmeter. The 500-μA-based voltmeter *did not* provide an accurate reading for the circuit in Fig. 4-18A since R_{in1} (20 kΩ) was *not* large compared to R_{TH} (20 kΩ). Remember, it is the responsibility of the person making a measurement to select an appropriate instrument for that measurement.

4-7 REVIEW OF OBJECTIVES

A basic meter movement can be converted into a voltmeter with the addition of a series (multiplier) resistor. The multiplier is chosen so that full-scale deflection of the basic meter movement occurs when V_{FS} is placed across the series combination of the multiplier resistor and meter movement.

Multiple-range voltmeters often employ a series arrangement of multiplier resistors. This permits standard resistance values to be used for all but one of the multiplier resistors. Voltmeter sensitivity is defined as $S = 1/I_{FS}$ and has the dimensions of ohms per volt. Voltmeter sensitivity enables you to quickly calculate the input resistance (R_{in}) of a voltmeter since $R_{in} = SV_{FS}$.

Voltmeter loading effects were discussed in detail. In addition you learned how to design basic voltmeters and predict the percent accuracy and loading error. Numerous guidelines were provided to minimize errors. Specifically, if $R_{in} \gg R_{TH}$ the loading error will not be significant. As was the case with ammeters, calibration error is minimized by selecting a range that provides up-scale readings.

4-8 QUESTIONS

1. Is it desirable for a voltmeter to have a large or small value of R_{in}? Why?
2. What does the input resistance of a voltmeter depend on?
3. Why are separate jacks normally provided for low- and high-voltage ranges?
4. What is the function of D_1 and D_2 in Fig. 4-7?
5. What is the sensitivity of a voltmeter employing a 1-mA, 100-Ω meter movement on the 10-V range? The 50-V range?

4-9 PROBLEMS

1. What series resistance is required for the ranges 10 V, 50 V, 250 V, and 1000 V if a 1-mA, 100-Ω meter movement is employed in the design of the voltmeter?
2. What is the sensitivity of the voltmeter in Problem 1? What is the input resistance of each range?
3. Sketch (with values) a series arrangement of multiplier resistors for the voltmeter in Problem 1.
4. The potentiometer in Fig. 4-20A is adjusted so that $R_{AB} = 200\ \Omega$. What is

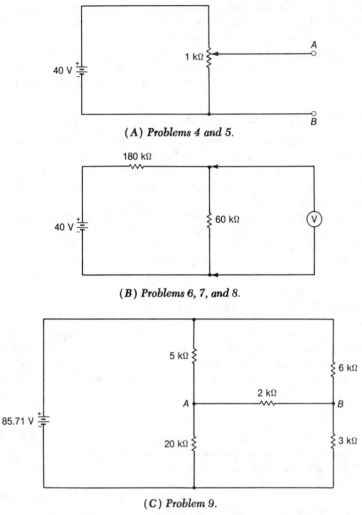

(A) Problems 4 and 5.

(B) Problems 6, 7, and 8.

(C) Problem 9.

Fig. 4-20. Schematic diagrams for Chapter 4 problems.

the voltage from A to B? What will a voltmeter on the 10-V range read when connected across AB if it employs a 1-mA meter movement?

5. Dr. Courtine recommends changing the 1-kΩ potentiometer in Fig. 4-20A to 1 MΩ in order to "reduce the current drain on the 40-V power supply." Assuming the change is made and the potentiometer has been adjusted so that $R_{AB} = 200$ kΩ, what is the voltage from A to B? What will the voltmeter in Problem 4 read?

6. The voltage across the 60-kΩ resistor in Fig. 4-20B is measured with a voltmeter whose sensitivity is 1000 Ω/V. Determine the percent accuracy and loading error assuming the 50-V range is employed for the measurement.

7. Repeat Problem 6 assuming the sensitivity of the voltmeter is 20 kΩ/V.

8. For a 3-percent calibration error predict the range within which the measured voltage in Problem 7 should lie.

9. Determine the Thevenin equivalent circuit for the circuit shown in Fig. 4-20C. For the voltage between terminals A and B to be measured with 1-percent loading error or less, what is the smallest value R_{in} can have?

4-10 EXPERIMENT 4-1

Objective

The objective of this experiment is to design a multiple-range voltmeter employing a series arrangement of multiplier resistors.

Material Required

50-μA meter movement
9-V battery
1-kΩ and 5-kΩ potentiometers
Resistor or resistor decades (you will calculate the required values as part of the experiment)
Oscilloscope

Introduction

To convert a basic meter movement into a voltmeter, series (multiplier) resistors are employed. Equations 4-5 and 4-6 are used to determine the value of R_s. Specifically,

$$R_{in} = SV_{FS} \qquad (4\text{-}5)$$

$$R_s = SV_{FS} - R_m \qquad (4\text{-}6)$$

Frequently a variable resistor is employed to compensate for variations in the internal resistance of meter movements, component values, etc.

Procedure

Step 1. Assume R_{CAL} in Fig. 4-21 has been adjusted so that $R_{CAL} + R_m = 8$ kΩ.

Step 2. Use Equation 4-5 to calculate R_{in} for the following ranges: 5 V, 10 V, and 50 V.

5-V range: $R_{in} =$ _____

10-V range: $R_{in} =$ _____

50-V range: $R_{in} =$ _____

Step 3. Use Equation 4-6 to calculate the required value of R_s for each range. Since $R_{CAL} + R_m = 8$ kΩ, use 8 kΩ for R_m in Equation 4-6.

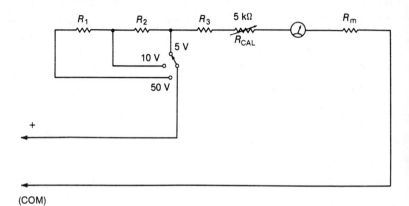

Fig. 4-21. Multiple-range voltmeter for Experiment 4-1.

5-V range: $R_s =$ _____

10-V range: $R_s =$ _____

50-V range: $R_s =$ _____

Step 4. Determine the values required for R_1, R_2, and R_3 in Fig. 4-21.

$R_1 =$ _____ $R_2 =$ _____

$R_3 =$ _____

Step 5. Build the voltmeter shown in Fig. 4-21. Adjust R_{CAL} so that $R_{CAL} + R_m = 8$ kΩ.

Step 6. Adjust the 1-kΩ potentiometer in Fig. 4-22 so that $R_{AB} = 0$. Hook the voltmeter to terminals A and B.

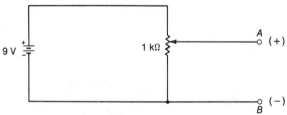

Fig. 4-22. Test circuit of Experiment 4-1.

Step 7. Set the voltmeter to the 5-V range. With an oscilloscope connected across A and B carefully adjust the 1-kΩ potentiometer until $V_{AB} = 5$ V. You should observe full-scale deflection of the voltmeter. If necessary adjust R_{CAL} to obtain full-scale deflection.

Step 8. Switch to the 10-V and 50-V ranges. Record the voltmeter reading in each case.

        10-V range: $V =$ _____

        50-V range: $V =$ _____

Step 9. Adjust V_{AB} to 9 V and record the voltmeter readings on the 10-V and 50-V ranges.

        10-V range: $V =$ _____

        50-V range: $V =$ _____

Step 10. Assume a 3-percent calibration error. Do all voltmeter readings fall within the expected ranges?

Discussion

Resistor R_{CAL} can be used to individually calibrate each range. If precision (1-percent, 0.1-percent, etc.) resistors are used for the multiplier resistors, very little calibration is necessary.

If you modify the voltmeter in Fig. 4-21 to include a low-voltage and high-voltage range, separate jacks should be provided for these ranges. A modest overload protection can be obtained by employing diodes across the meter movement as shown in Fig. 4-7.

Conclusions

What are sources of error in this experiment? How can they be minimized? Why would a 50-μA movement be preferred to a 1-mA movement in a voltmeter?

4-11 EXPERIMENT 4-2

Objective

The objective of this experiment is to investigate voltmeter loading effects using the voltmeter designed in Experiment 4-1.

Material Required

The voltmeter you designed in Experiment 4-1
9-V battery
Resistors: 1 kΩ, 10 kΩ, 100 kΩ, and 1 MΩ (two of each)
Oscilloscope

Introduction

Voltmeter accuracy a and loading error e are defined by Equations 4-9 and 4-11, respectively. Specifically,

$$a = \frac{R_{\text{in}}}{R_{\text{in}} + R_{TH}} \qquad (4\text{-}9)$$

$$e = 1 - a \qquad (4\text{-}11)$$

Good accuracy results when the condition that $R_{in} \gg R_{TH}$ is satisfied. If R_{in} is not large compared to R_{TH}, then significant loading error will be present and the measured values will be inaccurate.

Procedure

Step 1. Calculate R_{in} for the 5-V range. Enter your result in Table 4-6.

Step 2. Calculate R_{TH} for $R = 1$ kΩ, 10 kΩ, 100 kΩ, and 1 MΩ in Fig. 4-23. Enter the values in Table 4-6.

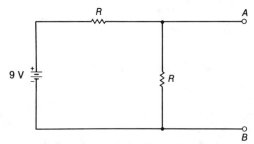

Fig. 4-23. Test circuit of Experiment 4-2.

Step 3. Use Equation 4-9 to estimate the accuracy and measured value for V_{AB} in Fig. 4-23 when $R = 1$ kΩ, 10 kΩ, 100 kΩ, and 1 MΩ. Record your results in Table 4-6.

Step 4. Build the circuit and make all appropriate measurements.

Step 5. Repeat Steps 1–4 for the 0- to 10-V and 0- to 50-V ranges.

Step 6. Compare calculated and measured values.

Discussion

Your calculated and measured values should be in close agreement. You should check to see if your measured values are within the ranges expected when the effects of calibration error are taken into account. Additional sources of error include resistor tolerances and variations in battery voltage from one 9-V battery to the next. By measuring your battery voltage with an oscilloscope when it is in the circuit, you can take such variations into account when you estimate (Step 3) the measured value of V_{AB}.

Conclusions

When should voltmeter loading error be considered? When you are using a voltmeter what is a good indication that significant loading error is present? What can be done to minimize loading error?

Table 4-6. Calculated and Measured Values for Experiment 4-2

Quantity	Resistance R			
	1 kΩ	10 kΩ	100 kΩ	1 MΩ
5-V range ($R_{in} =$)				
R_{TH}				
a				
V_{AB} (estimated)				
V_{AB} (measured)				
10-V range ($R_{in} =$)				
R_{TH}				
a				
V_{AB} (estimated)				
V_{AB} (measured)				
50-V range ($R_{in} =$)				
R_{TH}				
a				
V_{AB} (estimated)				
V_{AB} (measured)				

CHAPTER 5

Ohmmeters

5-1 INTRODUCTION

The basic meter movement is a versatile device. You have seen how a wide range of dc currents and voltages can be accurately measured using a basic meter movement and a few additional components. In this chapter the techniques used to measure resistance will be discussed. As you will see, there are two basic types of ohmmeters —the *series* ohmmeter and the *shunt* ohmmeter. The principles associated with each type of ohmmeter, and examples illustrating the design of elementary ohmmeters, are provided in the following sections. In addition, you will become familiar with the operation of a *voltmeter*-type ohmmeter. The voltmeter-type ohmmeter is employed in many commercial instruments.

5-2 OBJECTIVES

At the end of this chapter you will be able to do the following:

• Explain how series-, shunt-, and voltmeter-type ohmmeters operate.
• Design elementary series- and shunt-type ohmmeters.
• Explain the function of series and shunt OHMS ADJUST controls.
• Design elementary multiple-range ohmmeters.
• Understand how various ohmmeter scales are calibrated.

5-3 THE SERIES OHMMETER

The basic series-type ohmmeter is illustrated in Fig. 5-1. Notice in Fig. 5-1 that the instrument consists of a battery (V), a basic meter

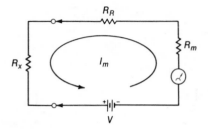

Fig. 5-1. A series-type ohmmeter.

movement, and a series resistor (R_R). The resistance you wish to measure (R_x) is connected in *series* with the ohmmeter. As you will see, the range of resistance values that can be measured with this ohmmeter is determined by the voltage source V and the value of the *range resistor* R_R.

The basic idea behind the operation of this circuit is to obtain a current proportional to the unknown resistance (R_x) and then measure that current with the basic meter movement. Naturally, the meter face will be calibrated to read resistance rather than current. With reference to Fig. 5-1, the meter face has to be calibrated to measure resistance, and the relationships necessary to design an elementary series-type ohmmeter have to be determined.

The resistance you wish to measure (R_x) obviously lies within the range from 0 (a short circuit) to ∞ (an open circuit). When $R_x = \infty$, $I_m = 0$ and no deflection of the basic meter movement occurs. To begin calibrating the meter face, ∞ is marked on the *left* side of the meter face, as illustrated in Fig. 5-2.

When $R_x = 0$, we want $I_m = I_{FS}$, so that full-scale deflection of the basic meter movement occurs. Therefore

$$I_m = \frac{V}{R_R + R_m} = \frac{V}{R_T} = I_{FS}$$

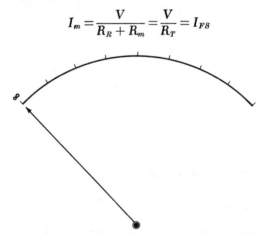

Fig. 5-2. Measuring $R_s = \infty$ for the series-type ohmmeter.

where R_T is simply $R_R + R_m$. Thus, in the calibration process, 0 must be marked on the right side of the meter face as illustrated in Fig. 5-3.

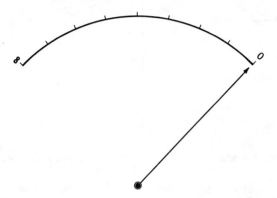

Fig. 5-3. Measuring $R_x = 0$ for the series-type ohmmeter.

When R_x is some finite value between 0 and ∞,

$$I_m = \frac{V}{R_x + R_T}$$

The resulting deflection (D) is therefore

$$D = \frac{I_m}{I_{FS}} = \frac{V}{R_x + R_T} \div \frac{V}{R_T} = \frac{V}{R_x + R_T} \times \frac{R_T}{V}$$

$$D = \frac{R_T}{R_x + R_T} \tag{5-1}$$

Solving Equation 5-1 for R_x yields

$$DR_x + DR_T = R_T$$
$$DR_x = R_T - DR_T$$
$$= R_T(1 - D)$$

Thus

$$R_x = \frac{R_T(1 - D)}{D} \tag{5-2}$$

Equation 5-2 is important. It tells you how the meter face should be calibrated in terms of R_T. Employing Equation 5-2 a table of values for percent deflection and R_x can be generated (Table 5-1). The calibrated meter face is illustrated in Fig. 5-4. Notice in Fig. 5-4 that the ohms scale is *nonlinear* and *"backwards."* The nonlinear ohms scale makes it difficult to read resistance values that produce small

105

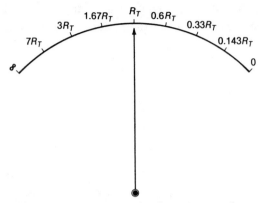

Fig. 5-4. Measuring $R_x = R_T$ for the series-type ohmmeter.

deflections, that is, large values of R_x. The "backwards" scale (zero ohms on the right) is *characteristic* of a series-type ohmmeter.

EXAMPLE 5-1

Design a series-type ohmmeter employing a standard 1.5-V D cell for the voltage source. As a change of pace, a 100-μA, 1-kΩ meter movement will be utilized in the design. Indicate how the ohms scale should be calibrated.

Table 5-1. Data to Calibrate the Series-Type Ohmmeter
(Fig. 5-4)

Percent Deflection ($D \times 100$)	$R_x = R_T(1 - D)/D$
0	$\infty \ \Omega$
12.5	$7R_T$
25	$3R_T$
37.5	$1.67R_T$
50	R_T
62.5	$0.6R_T$
75	$0.33R_T$
87.5	$0.143R_T$
100	$0 \ \Omega$

The total resistance of the meter is

$$R_T = \frac{V}{I_{FS}} = \frac{1.5 \text{ V}}{100 \ \mu\text{A}} = 15 \text{ k}\Omega$$

and the range resistance is

$$R_R = R_T - R_m$$
$$= 15 \text{ k}\Omega - 1 \text{ k}\Omega$$
$$= 14 \text{ k}\Omega$$

For $D = 0\%$,	$R_x = \infty$
For $D = 25\%$,	$R_x = 3(15\,\text{k}\Omega) = 45\,\text{k}\Omega$
For $D = 50\%$,	$R_x = 15\,\text{k}\Omega$
For $D = 75\%$,	$R_x = (1/3)(15\,\text{k}\Omega) = 5\,\text{k}\Omega$
For $D = 100\%$	$R_x = 0$

Figs. 5-5 and 5-6 illustrate the design and partially calibrated ohms scale.

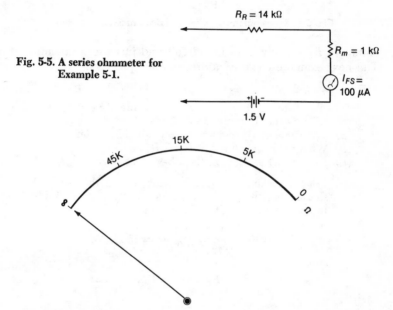

Fig. 5-5. A series ohmmeter for Example 5-1.

Fig. 5-6. Partially calibrated ohms scale for the series ohmmeter in Example 5-1.

5-4 OHMS ADJUST CONTROL

The elementary ohmmeter designed in Example 5-1 will work as "advertised" *if* the actual component values have the *exact* values indicated in the solution. In practice, this is rarely the case. You will remember from Experiment 2-2 that the terminal voltages of "real" batteries vary slightly from one battery to another. In addition, the terminal voltage of a given battery does *not* remain constant over long periods. A battery (like people) will "age" with time—its internal resistance increases, which reduces the available terminal voltage under load. To compensate for such changes you can employ the circuit illustrated in Fig. 5-7. The purpose of R_2 in Fig. 5-7 is to compensate for battery aging. Frequently R_2 is called the OHMS ADJUST or ZERO OHMS control. Perhaps you are wondering why we didn't

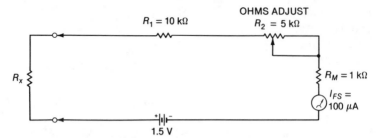

Fig. 5-7. Series ohmmeter with a series OHMS ADJUST control.

eliminate R_1 entirely from the circuit and just use a potentiometer. The next example cleverly addresses this question.

EXAMPLE 5-2

The ohmmeter in Fig. 5-7 employs a brand-new, "just-off-the-shelf," 1.5-V D cell. Determine:

(a) The value of R_2 required to zero the ohmmeter.
(b) What the ohmmeter will read if $R_x = 15$ kΩ.
(c) Repeat (a) and (b) if the 1.5-V D cell ages down to 1.1 V.

(a) Prior to making a resistance measurement the test leads are shorted together, and the OHMS ADJUST control is adjusted to read zero ohms. Thus $R_x = 0$ and

$$R_T = \frac{V}{I_{FS}} = \frac{1.5 \text{ V}}{100 \text{ }\mu\text{A}} = 15 \text{ k}\Omega$$

Also,

$$R_T = R_1 + R_2 + R_m$$

so that

$$
\begin{aligned}
R_2 &= R_T - (R_1 + R_m) \\
&= 15 \text{ k}\Omega - (10 \text{ k}\Omega + 1 \text{ k}\Omega) \\
&= 15 \text{ k}\Omega - 11 \text{ k}\Omega \\
&= 4 \text{ k}\Omega
\end{aligned}
$$

(b)

$$
D = \frac{R_T}{R_x + R_T} = \frac{15 \text{ k}\Omega}{15 \text{ k}\Omega + 15 \text{ k}\Omega} = \frac{15 \text{ k}\Omega}{30 \text{ k}\Omega}
$$
$$
= 0.5 = 50\%
$$

Table 5-1 indicates that the ohmmeter will read 15 kΩ.

(c) Again, prior to measuring R_x the test leads are first shorted and R_2 adjusted to zero the ohmmeter. Thus

$$R_T = \frac{V}{I_{FS}} = \frac{1.1 \text{ V}}{100 \text{ }\mu\text{A}} = 11 \text{ k}\Omega$$

$$R_2 = R_T - (R_1 + R_m)$$

$$R_2 = 11\,k\Omega - (10\,k\Omega + 1\,k\Omega)$$
$$= 11\,k\Omega - 11\,k\Omega = 0\,\Omega$$
$$D = \frac{R_T}{R_x + R_T} = \frac{11\,k\Omega}{15\,k\Omega + 11\,k\Omega} = \frac{11\,k\Omega}{26\,k\Omega} = 0.423$$
$$= 42.3\,\%$$

The meter face *has been calibrated* based on the assumption that $V = 1.5$ V and $R_T = 15$ kΩ. Due to the fact that the battery has aged down to 1.1 V, a 15-kΩ resistor for R_x will only produce a deflection of 42.3 percent. Thus the ohmmeter will *read* the value of R_x in Fig. 5-6 that corresponds to a deflection of 42.3 percent. Therefore

$$R_x = \frac{R_T(1 - D)}{D}$$
$$= \frac{(15\,k\Omega)(1 - 0.423)}{0.423} = 20.46\,k\Omega$$

Thus R_2 enables you to compensate for battery aging *at the expense* of introducing an error in your measurement. To limit this error to a reasonable value, the adjustable range of R_2 is limited. This is the reason why we did not eliminate R_1 in Fig. 5-7, in favor of a potentiometer. If the battery voltage in Example 5-2 decreased below 1.1 V, it would no longer be possible to zero the ohmmeter. When this condition occurs it *indicates that the battery should be replaced.* If the amount of error illustrated in Example 5-2 is considered excessive, you could further limit the adjustable range of R_2. The disadvantage of this solution, of course, is that the range of compensation for battery aging is also reduced. A much better solution is illustrated in Fig. 5-8 where the OHMS ADJUST control has been placed in parallel with the meter movement. In Example 5-3 the superiority of this design is proved.

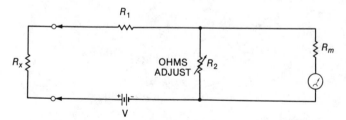

Fig. 5-8. Series ohmmeter with a shunt OHMS ADJUST control.

EXAMPLE 5-3

Your old, reliable, 50-μA, 2-kΩ meter movement is used in the circuit shown in Fig. 5-8, where $R_1 = 14$ kΩ and $V = 1.5$ V. Determine:

(a) The value of R_2 required to zero the ohmmeter.

(b) How the meter face should be calibrated.

(c) What the ohmmeter will read if $R_x = 15$ kΩ.

(d) The value of R_2 required to zero the ohmmeter if the battery voltage drops to 1.1 V.

(e) What the ohmmeter will indicate in (d).

(a) To zero the ohmmeter the test leads are first shorted together making $R_x = 0$. Resistor R_2 (the OHMS ADJUST control) is then varied to obtain a full-scale deflection. In order to calculate the value required for R_2, we will determine the current through R_2 and voltage across R_2 when full-scale deflection occurs. Thus, in Fig. 5-9,

$$V_{R2} = V_{Rm}$$
$$= I_{FS}R_m$$
$$= (50 \ \mu A)(2 \ k\Omega)$$
$$= 0.1 \ V$$

$$V_{R1} = V - V_{R2}$$
$$= 1.5 \ V - 0.1 \ V$$
$$= 1.4 \ V$$

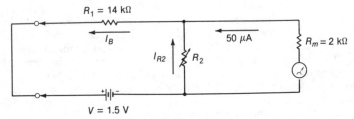

Fig. 5-9. Zeroing the ohmmeter in Example 5-3 when $V = 1.5$ V.

The battery current is

$$I_B = I_{R1} = \frac{V_{R1}}{R_1} = \frac{1.4 \ V}{14 \ k\Omega} = 100 \ \mu A$$

$$I_{R2} = I_B - I_m$$
$$= 100 \ \mu A - 50 \ \mu A = 50 \ \mu A$$

Therefore

$$R_2 = \frac{V_{R2}}{I_{R2}} = \frac{0.1 \ V}{50 \ \mu A} = 2 \ k\Omega$$

(b) Recall that a shunted meter movement can be converted to its series equivalent. This conversion is illustrated in Fig. 5-10. Fig. 5-11 represents the resulting equivalent circuit. Fig. 5-11 should look familiar! It is identical with Fig. 5-5,

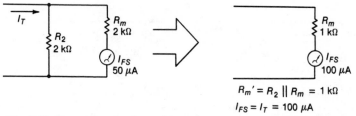

Fig. 5-10. Converting the shunted meter movement of Example 5-3 to its series equivalent.

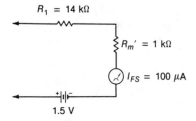

Fig. 5-11. Series equivalent circuit for Example 5-3.

which is for the ohmmeter designed in Example 5-1. Thus the calibrated meter face will be identical with Fig. 5-6, which is reproduced in Fig. 5-12.

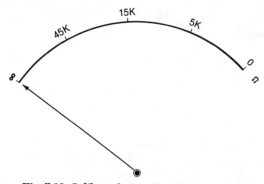

Fig. 5-12. Calibrated meter face for Example 5-3.

(c) From the series equivalent circuit in Fig. 5-11 it is obvious that when $R_x = 15$ kΩ, half-scale deflection results. Thus the ohmmeter will read 15 kΩ.

(d) Prior to measuring resistance, the test leads are shorted ($R_x = 0$) and R_2 is adjusted to obtain full-scale deflection. This is illustrated in Fig. 5-13. Thus

$$V_{R2} = V_{Rm}$$
$$= I_{FS}R_m$$

111

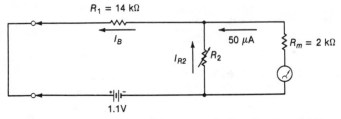

Fig. 5-13. Zeroing the ohmmeter in Example 5-3 when V = 1.1 V.

$$V_{R2} = (50 \ \mu A)(2 \ k\Omega)$$
$$= 0.1 \ V$$

and

$$V_{R1} = V - V_{R2}$$
$$= 1.1 \ V - 0.1 \ V$$
$$= 1 \ V$$

Also,

$$I_B = I_{R1} = \frac{V_{R1}}{R_1} = \frac{1 \ V}{14 \ k\Omega} = 71.43 \ \mu A$$

and

$$I_{R2} = I_B - I_m$$
$$= 71.43 \ \mu A - 50 \ \mu A = 21.43 \ \mu A$$

Therefore

$$R_2 = \frac{V_{R2}}{I_{R2}} = \frac{0.1 \ V}{21.43 \ \mu A} = 4.67 \ k\Omega$$

(e) First we will calculate the value of I_m in Fig. 5-14.

$$4.67 \ k\Omega \parallel 2 \ k\Omega = \frac{4.67 \times 2}{4.67 + 2} = \frac{9.34}{6.67} = 1.4 \ k\Omega$$

The voltage across the equivalent 1.4-kΩ resistor equals the voltage across R_m. This voltage can be determined using the voltage division principle:

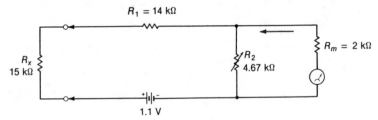

Fig. 5-14. Measuring a 15-kΩ resistor for V = 1.1 V with the ohmmeter in Example 5-3.

$$V_{1.4k\Omega} = V_{Rm} = \frac{(1.1\,\text{V})\,(1.4\,\text{k}\Omega)}{30.4\,\text{k}\Omega} = 50.66\,\text{mV}$$

Therefore

$$I_m = \frac{50.66\,\text{mV}}{2\,\text{k}\Omega} = 25.33\,\mu\text{A}$$

Knowing the value of I_m enables us to determine the percent deflection:

$$D = \frac{I_m}{I_{FS}} = \frac{25.33\,\mu\text{A}}{50\,\mu\text{A}} = 0.507 = 50.7\%$$

The ohmmeter will read the value of R_x that corresponds to a deflection of 50.7 percent. Thus

$$\begin{aligned}
R_x &= \frac{R_T(1-D)}{D} \\
&= \frac{(15\,\text{k}\Omega)\,(1-0.507)}{0.507} \\
&= 14.59\,\text{k}\Omega
\end{aligned}$$

Comparing the results of Examples 5-2 and 5-3 you can see the advantage of a shunt OHMS ADJUST control.

5-5 THE SHUNT OHMMETER

An elementary shunt-type ohmmeter is illustrated in Fig. 5-15. Notice that the resistance to be measured (R_x) is placed in parallel

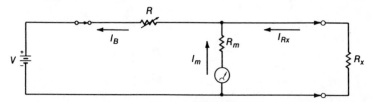

Fig. 5-15. A shunt-type ohmmeter.

(shunt) with the basic meter movement. The shunt ohmmeter is similar to the series ohmmeter in one important respect. In either case the current through the meter movement (I_m) is proportional to the resistance being measured (R_x). In order to calibrate the shunt ohmmeter a number of working relationships need to be derived.

When $R_x = \infty$ in Fig. 5-15, R is adjusted so that a full-scale deflection of the basic meter occurs. To begin calibrating the meter face, ∞ is marked on the *right* side of the meter face as illustrated in Fig. 5-16. This is the *opposite* of the series ohmmeter.

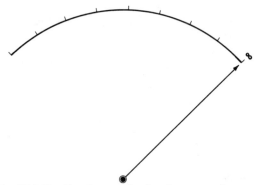

Fig. 5-16. Reading $R_x = \infty$ for the shunt-type ohmmeter.

When $R_x = 0$ the meter movement is shunted with a short circuit. This means that all of the battery current (I_B) flows through the short. Since no current flows through the meter movement, $I_m = 0$ and no deflection of the basic meter movement occurs. To continue calibrating the meter face you mark 0 on the *left* side of the meter face as illustrated in Fig. 5-17. To continue the calibration process, an equation should be derived that relates R_x to the amount of deflection (D):

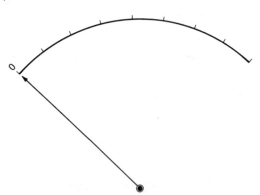

Fig. 5-17. Reading $R_x = 0$ for the shunt-type ohmmeter.

When $R_x = \infty$, R is adjusted so that $I_B = I_m = I_{FS}$. Thus

$$I_{FS} = \frac{V}{R + R_m}$$

When R_x is some finite value between 0 and ∞,

$$I_B = \frac{V}{R + R_m \| R_x} = \frac{V}{R + R_m R_x / (R_m + R_x)}$$

$$I_B = \frac{V(R_m + R_x)}{R(R_m + R_x) + R_m R_x}.$$

The battery current (I_B) divides between R_m and R_x. The portion of I_B that flows through the meter movement can be found by the current division principle. Therefore

$$
\begin{aligned}
I_m &= I_B \left(\frac{R_x}{R_m + R_x} \right) \\
&= \frac{V(R_m + R_x)}{R(R_m + R_x) + R_m R_x} \times \frac{R_x}{R_m + R_x} \\
&= \frac{V R_x}{R(R_m + R_x) + R_m R_x}
\end{aligned}
$$

Now

$$
\begin{aligned}
D &= \frac{I_m}{I_{FS}} = \frac{V R_x}{R(R_m + R_x) + R_m R_x} \div \frac{V}{R + R_m} \\
&= \frac{V R_x}{R(R_m + R_x) + R_m R_x} \times \frac{R + R_m}{V} \\
&= \frac{R_x (R + R_m)}{R(R_m + R_x) + R_m R_x} \\
&= \frac{R_x (R + R_m)}{R R_m + R R_x + R_m R_x} \\
&= \frac{R_x (R + R_m)}{R_x (R + R_m) + R R_m}
\end{aligned}
$$

Dividing numerator and denominator by $R + R_m$ yields

$$D = \frac{R_x}{R_x + R R_m / (R + R_m)}$$

The right-hand term in the denominator represents the parallel combination of R and R_m. We will define this as R_p. Thus

$$R_p = R \parallel R_m = \frac{R R_m}{R + R_m} \tag{5-3}$$

$$D = \frac{R_x}{R_x + R_p} \tag{5-4}$$

Solving Equation 5-4 for R_x,

$$
\begin{aligned}
D R_x + D R_p &= R_x \\
R_x - D R_x &= D R_p \\
R_x (1 - D) &= D R_p
\end{aligned}
$$

Therefore

$$R_x = \frac{DR_p}{1 - D} \qquad (5\text{-}5)$$

Equation 5-5 is analogous to Equation 5-2. It tells you how to calibrate the meter face of a shunt ohmmeter in terms of R_p. Employing Equation 5-5 we generate Table 5-2 and illustrate the calibrated meter face in Fig. 5-18. Notice that the ohms scale for the shunt-type ohmmeter is *not* backwards. Remember, a backwards scale is characteristic of a series-type ohmmeter.

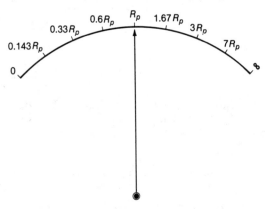

Fig. 5-18. Reading $R_x = R_p$ for the shunt-type ohmmeter.

Table 5-2. Data to Calibrate the Shunt-Type Ohmmeter
(Fig. 5-18)

Percent Deflection ($D \times 100$)	$R_x = DR_p/(1 - D)$
0	$0\ \Omega$
12.5	$0.143R_p$
25	$0.333R_p$
37.5	$0.6R_p$
50	R_p
62.5	$1.67R_p$
75	$3R_p$
87.5	$7R_p$
100	$\infty\ \Omega$

EXAMPLE 5-4

Design a shunt-type ohmmeter employing a 100-μA, 1-kΩ meter movement. The battery for the ohmmeter is assumed to be 1.5 V. Indicate how the ohms scale should be calibrated. Referring to Fig. 5-15, when $R_x = \infty$, R is adjusted so that $I_B = I_m = I_{FS}$. Thus

$$R + R_m = \frac{V}{I_{FS}} = \frac{1.5\,\text{V}}{100\,\mu\text{A}} = 15\,\text{k}\Omega$$

$$R = 15\,\text{k}\Omega - R_m = 15\,\text{k}\Omega - 1\,\text{k}\Omega = 14\,\text{k}\Omega$$

To calibrate the ohms scale we need to determine the value of R_p:

$$R_p = R \,\|\, R_m = \frac{(14\,\text{k}\Omega)(1\,\text{k}\Omega)}{15\,\text{k}\Omega} = 933.3\,\Omega$$

For $D = 0\%$, $R_x = 0\,\Omega$

For $D = 25\%$, $R_x = 0.333(933.3) = 310.8\,\Omega$

For $D = 50\%$, $R_x = 933.3\,\Omega$

For $D = 75\%$, $R_x = 3(933.3) = 2\,800\,\Omega$

For $D = 100\%$, $R_x = \infty\,\Omega$

For values of R_x expressed in terms of R_p, see Table 5-2.

Fig. 5-19 illustrates the design, and Fig. 5-20 contains the partially calibrated ohms scale. From Fig. 5-20 you can see that the

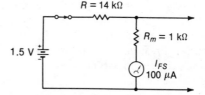

Fig. 5-19. A shunt ohmmeter for Example 5-4.

shunt-type ohmmeter is especially suited for the measurement of *small* resistance values. In order to increase the life of the battery in Fig. 5-19, you should add an ON/OFF switch in series with the battery. This will reduce the battery current to zero when the instrument is not being used (assuming you remember to shut the instrument off!). In addition, it would be desirable to add an OHMS ADJUST control in order to compensate for battery aging. Since the

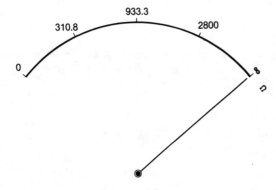

Fig. 5-20. Partially calibrated ohms scale for the shunt ohmmeter in Example 5-4.

shunt-type ohmmeter is *not* as popular as the series-type ohmmeter we will not discuss it further. At this point we will return to the series ohmmeter and learn how to add multiple-resistance ranges.

5-6 MULTIPLE RANGES

An ohmmeter scale is nonlinear. In a series-type ohmmeter, ∞ corresponds to zero deflection. Thus, resistance values which produce small deflections (large values of R_x) will be difficult to read. In practical ohmmeters, multiple resistance ranges are normally provided so that a wide range of resistances can be easily measured. The term "range" is used loosely here, since technically speaking the ranges of the ohmmeters discussed previously were infinite (∞ to 0 is ∞). An infinite range is of little use, if, in practice, it is difficult to accurately read the measured value!

A little thought will reveal the fact that calibration error has an effect on resistance measurements in addition to the effects discussed previously for current and voltage measurements. An ammeter (or voltmeter) has a linear scale whose range is finite. Thus the error in a current (or voltage) measurement due to calibration error is a constant number of amperes (or volts) all along the entire scale. Since an ohmmeter has a nonlinear scale whose range is infinite, the error which arises from calibration error is *not* a constant number of ohms all along the scale. It can be shown that this error, expressed as a percentage of the measured value, is smallest at the midscale position. Thus, when we speak of multiple ranges for ohmmeters we mean that the various ranges have different resistance values corresponding to the midscale position. It is good measurement practice to select a range so that the measured value of resistance is near the midscale position. This measurement technique will minimize the effect of calibration error for resistance measurements.

5-7 HIGHER RANGES

When $R_x = R_T$, a half-scale deflection of the basic meter movement results. Thus, *to increase the range by a factor of n increase the value of V and R_T by the same factor.*

EXAMPLE 5-5

Modify the ohmmeter in Fig. 5-21A so that midscale deflection corresponds to 150 kΩ rather than 15 kΩ. Now

$$n = \frac{150 \text{ k}\Omega}{15 \text{ k}\Omega} = 10$$

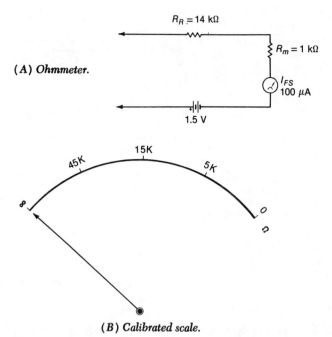

(A) Ohmmeter.

(B) Calibrated scale.

Fig. 5-21. A series ohmmeter and calibrated scale for Example 5-5.

Therefore

$$V = 10(1.5\,\text{V}) = 15\,\text{V}$$
$$R_T = 10(15\,\text{k}\Omega) = 150\,\text{k}\Omega$$
$$R_R = R_T - R_m$$
$$= 150\,\text{k}\Omega - 1\,\text{k}\Omega = 149\,\text{k}\Omega$$

Fig. 5-22 illustrates the modified ohmmeter and partially calibrated scale. The original and modified ohmmeters can easily be combined into a single instrument. This is illustrated in Fig. 5-23. In Fig. 5-23 an $R \times 1$ and $R \times 10$ range is provided, and a common scale is employed for each range.

5-8 LOWER RANGES

A series-type ohmmeter can be used to measure "small" values of resistance by *shunting the meter movement and range resistor* with a resistance (R) such that

$$R << R_m + R_R$$

This concept is illustrated in Example 5-6.

119

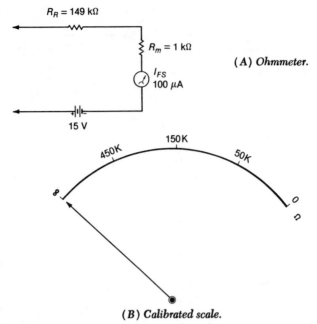

(A) Ohmmeter.

$R_R = 149\ k\Omega$

$R_m = 1\ k\Omega$

I_{FS} 100 μA

15 V

150K

450K   50K

8   0

(B) Calibrated scale.

Fig. 5-22. An ohmmeter and calibrated scale with a 150-kΩ midscale reading.

EXAMPLE 5-6

The meter movement and range resistor of Example 5-1 are shunted by a resistance of 15 Ω. This is illustrated in Fig. 5-24. Determine how the scale should be calibrated.

We will reduce the ohmmeter to its series equivalent. Thus

$$R_m' = 15\ \Omega\ ||\ 15\ k\Omega \cong 15\ \Omega$$

When $R_x = 0$,

$$I_{15\Omega} = \frac{1.5\ V}{15\ \Omega} = 0.1\ A$$

$$I_m = \frac{1.5\ V}{15\ k\Omega} = 100\ \mu A$$

Thus the effective I_{FS} of the shunted movement is

$$I_T = I_{15\Omega} + I_m$$
$$= 0.1\ A + 100\ \mu A$$
$$\cong 0.1\ A = I_{FS}'$$

The series equivalent circuit is illustrated in Fig. 5-25. Once the series equivalent circuit is obtained, it is a simple matter to calibrate the scale. This is accomplished by utilizing Table 5-1 and is illustrated in Fig. 5-26.

120

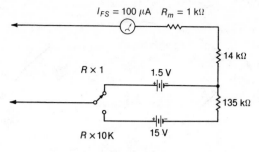

(A) Ohmmeter.

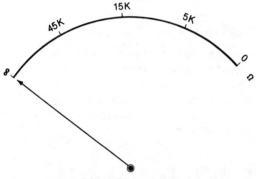

(B) Calibrated scale.

Fig. 5-23. Combining the ohmmeters in Fig. 5-21 and 5-22 into a
single instrument.

Fig. 5-24. A low-resistance ohmmeter
for Example 5-6.

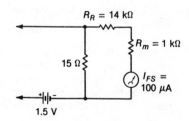

Fig. 5-25. Series equivalent for the
ohmmeter in Fig. 5-24.

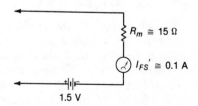

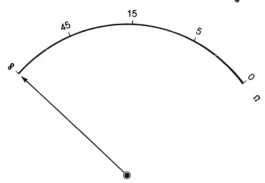

Fig. 5-26. Partially calibrated ohms scale for Example 5-6.

5-9 LIMITATIONS ON BATTERY SIZE

In Section 5-6 you saw how the range of a series-type ohmmeter could be increased by a factor of n simply by increasing V and R_T by the same factor. In commercial instruments n is some multiple of 10. Thus ohmmeter ranges of $R \times 1$, $R \times 10$, $R \times 100$, $R \times 1\,000$, and $R \times 10\,000$ are quite common. An advantage of this approach is that the *same scale* can be used for each range.

EXAMPLE 5-7

Assume a series-type ohmmeter employs a 1.5-V battery on the $R \times 1$ range. What voltages are required for the $R \times 100$- and $R \times 10\,000$-ohm ranges?

For the $R \times 100$-ohm range, $n = 100$, and therefore

$$V = 100(1.5\text{ V}) = 150\text{ V}$$

For the $R \times 10\,000$-ohm range, $n = 10\,000$, and therefore

$$V = 10\,000(1.5\text{ V}) = 15\text{ kV}!$$

Not only are 15-kV batteries difficult to obtain, they are quite impractical for portable equipment! Thus commercial ohmmeters utilize a different approach to obtain practical multiple-range ohmmeters. What is this approach? How does it work? The answers to these and other fascinating questions are provided in the next section.

5-10 THE VOLTMETER-TYPE OHMMETER

The ohmmeter illustrated in Fig. 5-27 is called a *voltmeter-type* ohmmeter. This is the type of ohmmeter typically found in commercial instruments. An advantage of the voltmeter-type ohmmeter is that large battery voltages are *not* required for high-resistance ranges. Notice in Fig. 5-27 that the voltmeter portion of the ohmmeter consists of the meter movement, R_1, and R_2. The idea is to

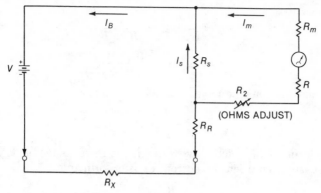

Fig. 5-27. Voltmeter-type ohmmeter.

obtain a voltage proportional to the unknown resistance (R_x) and then measure this voltage with the voltmeter portion of the ohmmeter. Naturally, the scale is calibrated to read ohms rather than volts. In a manner similar to the analysis provided for the series- and shunt-type ohmmeters, we will now derive some useful relationships for the voltmeter-type ohmmeter.

When $R_x = \infty$ I_B, I_s, and I_m are all zero. Thus no deflection of the meter movement takes place. To begin the process of calibrating the scale, ∞ would be marked on the *left* side of the scale.

When $R_x = 0$, R_2 would be adjusted to obtain full-scale deflection. Thus zero ohms corresponds to a full-scale deflection, so 0 is marked on the *right* side of the scale. Referring to Fig. 5-27, let R_v be the input resistance of the voltmeter. By inspection,

$$R_v = R_m + R_1 + R_2$$

We will define the parallel combination of R_s and R_v to be R_p. Thus

$$R_p = R_s \,||\, R_v = \frac{R_s R_v}{R_s + R_v}$$

To prevent the voltmeter from loading the circuit, $R_v \gg R_s$. In any case, when $R_x = 0$,

$$I_B = \frac{V}{R_p + R_R}$$

The portion of I_B that flows through the meter movement (I_{FS}) is easily obtained via the current division principle:

$$I_{FS} = I_B \times \frac{R_s}{R_s + R_v}$$

$$I_{FS} = \frac{V}{R_p + R_R} \times \frac{R_s}{R_s + R_v}$$

$$= \frac{VR_s}{(R_p + R_R)(R_s + R_v)}$$

When R_x is some finite value between 0 and ∞,

$$I_B = \frac{V}{R_p + R_R + R_x}$$

The portion of I_B which flows through the meter movement (I_m) is again easily found via the current division principle:

$$I_m = I_B \times \frac{R_s}{R_s + R_v}$$

$$= \frac{V}{R_p + R_R + R_x} \times \frac{R_s}{R_s + R_v}$$

$$I_m = \frac{VR_s}{(R_p + R_R + R_x)(R_s + R_v)}$$

Recalling that $D = I_m/I_{FS}$ we have

$$D = \frac{VR_s}{(R_p + R_R + R_x)(R_s + R_v)} \div \frac{VR_s}{(R_p + R_R)(R_s + R_v)}$$

$$= \frac{VR_s}{(R_p + R_R + R_x)(R_s + R_v)} \times \frac{(R_p + R_R)(R_s + R_v)}{VR_s}$$

$$= \frac{R_p + R_R}{R_p + R_R + R_x} \tag{5-6}$$

Finally, solving Equation 5-6 for R_x yields

$$DR_p + DR_R + DR_x = R_p + R_R$$

$$DR_x = R_p + R_R - DR_R - DR_p$$

$$= R_p(1 - D) + R_R(1 - D)$$

$$= (R_p + R_R)(1 - D)$$

Therefore

$$R_x = \frac{(R_p + R_R)(1 - D)}{D} \tag{5-7}$$

As you might guess, Equation 5-7 is used to generate Table 5-3. The calibrated meter face for the voltmeter-type ohmmeter is illustrated in Fig. 5-28.

Our analysis of the series- and shunt-type ohmmeters given previously made the analysis of the voltmeter-type ohmmeter a "piece of cake." One of the ways we learn is to *compare* new knowledge with

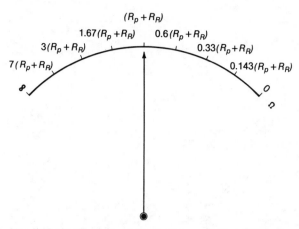

Fig. 5-28. For the voltmeter-type ohmmeter $R_x = R_p + R_R$.

**Table 5-3. Data to Calibrate the Voltmeter-Type Ohmmeter
(Fig. 5-28)**

Percent Deflection $(D \times 100)$	$R_x = (R_p + R_R)(1 - D)/D$
0	$\infty \ \Omega$
12.5	$7(R_p + R_R)$
25	$3(R_p + R_R)$
37.5	$1.67(R_p + R_R)$
50	$R_p + R_R$
62.5	$0.6(R_p + R_R)$
75	$0.33(R_p + R_R)$
87.5	$0.143(R_p + R_R)$
100	$0 \ \Omega$

old knowledge. Frequently generalizations are born from such comparisons. With this bit of philosophy in mind, we will compare our analysis of the series-type ohmmeter with the analysis just provided for the voltmeter-type ohmmeter. Specifically:

Series-Type Ohmmeter　　　　*Voltmeter-Type Ohmmeter*

$$D = \frac{R_T}{R_T + R_x} \quad (5\text{-}1) \qquad D = \frac{(R_p + R_R)}{(R_p + R_R) + R_x} \quad (5\text{-}6)$$

$$R_x = \frac{R_T(1 - D)}{D} \quad (5\text{-}2) \qquad R_x = \frac{(R_p + R_R)(1 - D)}{D} \quad (5\text{-}7)$$

The above comparison is meant to generate the desired *mental orgasm*, that is, the *generalization* that the voltmeter-type ohmmeter is in reality a cleverly disguised series-type ohmmeter! Fig. 5-29 illustrates the kind of ohmmeter typically found in commercial instru-

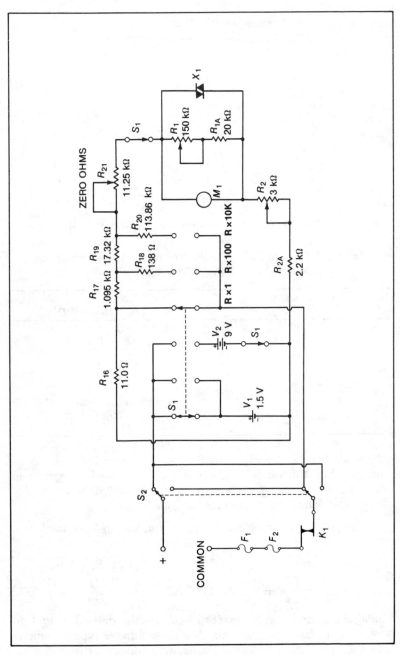

Fig. 5-29. Resistance ranges of the overload-protected Simpson 260–6 vom. (*Courtesy Simpson Electric Co.*)

ments. Notice that three resistance ranges are provided. The ohm-meter illustrated in Fig. 5-29 represents a practical design—that is, one that is a reasonable compromise between simplicity and performance.

EXAMPLE 5-8

For each range in Fig. 5-29, show that the circuit reduces to an equivalent voltmeter-type ohmmeter. In addition, indicate the value of R_R and R_s for each range.

Note in Fig. 5-29 that the meter movement (M_1) is shunted by two diodes. As you know, the diodes protect the meter movement in the event of an overload. When an overload condition does not exist the diodes act as open switches. Resistors R_1 and R_{1A} shunt the meter movement. Resistor R_1 would be adjusted to obtain the desired damping characteristics. Resistors R_2 and R_{2A} are connected in series with the meter movement to provide compensation for variations in component values, aging, etc. To simplify our analysis we will replace M_1, R_1, R_{1A}, R_2, R_{2A}, and the diodes by an equivalent meter movement. In addition, we will provide a simplified version of Fig. 5-29 which includes only those features necessary to reduce the circuit to an equivalent voltmeter-type ohmmeter.

Fig. 5-30 illustrates the simplified circuit. The circuit illustrated in Fig. 5-30 has been redrawn to indicate the values of R_R and R_s for each range in Figs. 5-31 $(R \times 1)$, 5-32 $(R \times 100)$, and 5-33 $(R \times 10K)$. If you compare these figures with Fig. 5-27 it is obvious that on each range the ohmmeter functions as a voltmeter-type ohmmeter. Table 5-4 summarizes the values of R_s and R_R for each range.

Fig. 5-34 illustrates a common use for an ohmmeter—checking continuity. Such checks will quickly reveal if a component is open or

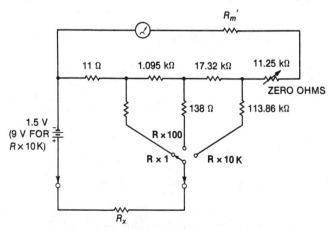

Fig. 5-30. Simplified Simpson 260–6 ohmmeter section.

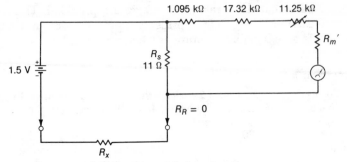

Fig. 5-31. Simplified R × 1 range.

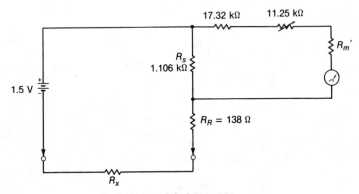

Fig. 5-32. Simplified R × 100 range.

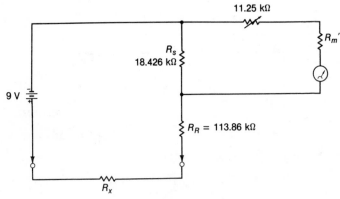

Fig. 5-33. Simplified R × 10K range.

shorted. The current that flows through the component being tested depends upon the resistance of the component *and the ohmmeter range employed for the measurement.* Table 5-5 provides the nomi-

Table 5-4. Summary of Resistance Values for the Multirange Ohmmeter (Example 5-8)

Range	R_s	R_B
R × 1	11 Ω	0 Ω
R × 100	1.106 kΩ	138 Ω
R × 10K	18.426 kΩ	113.86 kΩ

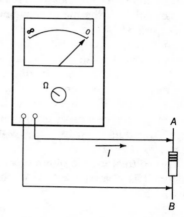

Fig. 5-34. Using an ohmmeter to check continuity.

nal short-circuit current for each resistance range of the Simpson 260® Series 6P and 6PM vom's. These data were obtained from the 260 operator's manual.

Table 5-5. Current for Different Resistance Ranges

Range	Current
R × 1	125 mA
R × 100	1.25 mA
R × 10K	75 μA

Notice in Table 5-5 that the battery current is quite large (125 mA) on the R×1 range. Also, notice that battery current decreases when higher-resistance ranges are employed. In Chapter 2 you learned that you should *not* measure the resistance of a meter movement directly with an ohmmeter. Even though the meter resistance would decrease the battery current, the current may still be large enough to damage the meter movement.

Often diodes are "checked out" with an ohmmeter. A *good* diode will have a small resistance when the ohmmeter is connected across it with one polarity, and a large resistance when the ohmmeter is connected with the opposite polarity. Table 5-5 indicates that you should employ a *high-resistance* range (so small currents are used)

129

for such a check. Naturally, the diode should be removed from the circuit for such a check.

5-11 REVIEW OF OBJECTIVES

Series ohmmeters have a characteristic "backwards" scale. An OHMS ADJUST (ZERO OHMS) control provides some compensation for battery aging in ohmmeters. This control can be connected in series or parallel with the meter movement. The parallel connection introduces less error than does the series connection. Shunt ohmmeters are not as widely used as series ohmmeters. Shunt ohmmeters are normally used to measure small resistance values.

Multiple ranges are very desirable for ohmmeters. The range of a series ohmmeter can be increased by increasing the values of V and R_T. Practical battery size limits the usefulness of this approach. Lower resistance ranges are obtained by shunting the meter movement and range resistor (R_R) with a resistance that is small compared to $R_m + R_R$.

The voltmeter-type ohmmeter is a special case of the series ohmmeter. It is a very practical circuit, commonly used in commercial ohmmeters.

5-12 QUESTIONS

1. What value of R_x can be read most accurately? Why?
2. What is meant by battery aging?
3. Should you use a low- or high-resistance range to make continuity checks? Why?
4. What is an advantage of a voltmeter-type ohmmeter over a standard series-type ohmmeter?
5. What is meant by a backwards scale?
6. Which type of ohmmeter has a backwards scale?
7. Why are deflection equations useful?
8. Dr. Courtine proposes modifying the ohmmeter in Fig. 5-24 so that "very small" resistance values can be measured. Specifically, the good doctor proposes replacing the 15-Ω resistor with a 0.1-Ω resistor. Critique the proposed modification. (Hint: examine the battery current required to zero the ohmmeter.)

5-13 PROBLEMS

1. The ohmmeter in Fig. 5-35A employs a 100-μA, 1-kΩ movement. Determine:
   (a) The value of R_R.
   (b) The values of R_x that produce deflections of 0, 25, 50, 75, and 100 percent.
   (c) The value of R_x corresponding to a deflection of 31.8 percent.
2. Assume the battery in Fig. 5-35A ages down to 2.5 V.

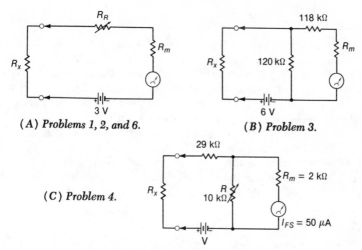

(A) Problems 1, 2, and 6.

(B) Problem 3.

(C) Problem 4.

Fig. 5-35. Schematic diagrams for Chapter 5 problems.

   (a) What value of R_R is required to zero the meter?
   (b) What resistance will the meter indicate when R_x is 30 kΩ? 10 kΩ?
3. The ohmmeter in Fig. 5-35B employs a 50-μA,2-kΩ movement.
   (a) Convert the ohmmeter to its series equivalent.
   (b) Make a neat sketch indicating how the meter face should be calibrated.
4. The value of V in Fig. 5-35C is 3 V.
   (a) What value of R is required to zero the meter?
   (b) If V ages down to 2.2 V, what value of R is required to zero the meter?
   (c) What will a 30-kΩ resistor read in (a) and (b)? A 10-kΩ resistor?
5. An unknown resistor (R_x) is measured with the ohmmeter in Fig. 5-19. The resulting deflection is 28 percent. What is the value of R_x?
6. What battery voltage would be required in Fig. 5-35A for an R × 10 range? What value of R_R would be required? Could you propose a "better" solution?

5-14 EXPERIMENT 5-1

Objective

The objective of this experiment is to construct an elementary series-type ohmmeter.

Material Required

50-μA meter movement
9-V battery
1-kΩ potentiometer
1-MΩ resistor decade, variable in 10-kΩ and 100-kΩ steps
10-kΩ resistor decade, variable in 1-kΩ and 100-kΩ steps
   If necessary, combinations of fixed resistors and potentiometers can be substituted for the resistor decades.
Oscilloscope or digital voltmeter (dvm)

Introduction

The series-type ohmmeter consists of a meter movement, battery (V), and range resistor (R_R) connected in series. The resistance to be measured (R_x) is connected in series with the instrument. By obtaining a current proportional to the unknown resistance, and measuring that current with the meter movement, it is possible to calibrate the meter face to read resistance directly.

The ohmmeter you will use in this experiment is illustrated in Fig. 5-36. Notice that the 1-kΩ potentiometer allows you to adjust the effective battery voltage (V) to any value between 0 V and 9 V. The following mathematical relationships along with Equations 5-1 and 5-2 are sufficient to analyze and design the ohmmeter. Specifically:

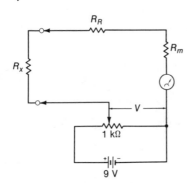

Fig. 5-36. Series-type ohmmeter for Experiment 5-1.

$$R_T = R_R + R_m = \frac{V}{I_{FS}}$$

$$R_R = R_T - R_m$$

$$D = \frac{R_T}{R_x + R_T} \tag{5-1}$$

$$R_x = \frac{R_T\,(1 - D)}{D} \tag{5-2}$$

Recall that Equation 5-2 was used to generate Table 5-1, which indicates how the meter face of a series-type ohmmeter should be calibrated.

Procedure

Step 1. Assume V in Fig. 5-36 to be 5 V. Calculate R_T for the ohmmeter.

$R_T =$ _____

Step 2. Calculate the value of R_R in Fig. 5-36.

$R_R =$ _____

Step 3. Build the ohmmeter circuit shown in Fig. 5-36. Adjust V to 5 V (an oscilloscope is recommended for this adjustment). Using Equation 5-2, calculate the values of R_x required to produce the deflections given in Table 5-6.

Step 4. Short the test leads of the ohmmeter ($R_x = 0$). If necessary adjust R_R to obtain full-scale deflection.

Step 5. Measure the deflections produced for *each* value of R_x in Table 5-6. Make a neat sketch of your calibrated meter face.

Step 6. Build the circuit shown in Fig. 5-37. Adjust V to 5 V. Short the test leads of the ohmmeter ($R_x = 0$) and zero the meter.

Step 7. Measure the deflection produced when $R_x = 100$ kΩ.
$D = $ _____ percent

Step 8. Remove R_x. To simulate a battery that has aged down to 4.5 V adjust the 1-kΩ potentiometer so that $V = 4.5$ V (an oscilloscope or dvm is recommended for this adjustment).

Step 9. Short the test leads of the ohmmeter ($R_x = 0$) and zero the meter.

Step 10. Measure the deflection produced when $R_x = 100$ kΩ. What value of R_x does the measured deflection correspond to based on your calibrated meter face?
$D = $ _____ percent
$R_x = $ _____

Conclusion

What factors affect the accuracy of a resistance measurement? What is the purpose of R in Fig. 5-37? If you wanted to add an R × 10 range, what battery voltage (V) would be required? List some

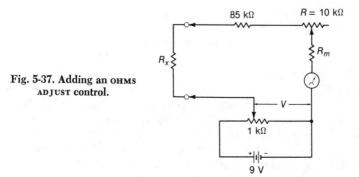

Fig. 5-37. Adding an OHMS ADJUST control.

advantages and disadvantages for the ohmmeter constructed in this experiment. Does the 1-kΩ potentiometer introduce a significant error into the experiment? Why?

Table 5-6. Data for Experiment 5-1

Deflection (D)	Percent Deflection	R_x	Percent Deflection (Measured)
0	0	$\infty \ \Omega$	
⅛	12.5		
¼	25		
⅜	37.5		
½	50		
⅝	62.5		
¾	75		
⅞	87.5		
1	100	$0 \ \Omega$	

5-15 EXPERIMENT 5-2

Objective

The objective of this experiment is to modify the ohmmeter of Experiment 5-1 to obtain a lower resistance range.

Material Required

50-μA meter movement
9-V battery
1-kΩ potentiometer
Resistor decades used in Experiment 5-1
Oscilloscope or digital voltmeter (dvm)

Introduction

A lower resistance range can be obtained with a series-type ohmmeter by shunting the meter movement and range resistor (R_R) with a resistance (R) such that $R \ll R_m + R_R$. Since $R_m + R_R$ in Experiment 5-1 equals approximately 100 kΩ, a 10-kΩ shunt resistance satisfies this relationship. This is illustrated in Fig. 5-38, which is the circuit that will be built in this experiment.

Procedure

Step 1. Assume that V in Fig. 5-38 is 5 V. Convert the ohmmeter to its *series* equivalent.

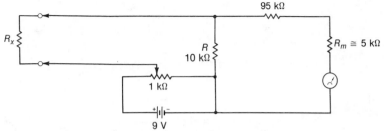

Fig. 5-38. A series-type ohmmeter for low-resistance measurements.

Step 2. Using Equation 5-2, calculate the values of R_x required to produce the deflections given in Table 5-7.

Step 3. Build the circuit shown in Fig. 5-38. Adjust V to 5 V (oscilloscope or dvm recommended for this adjustment).

Step 4. Short the test leads of the ohmmeter ($R_x = 0$). If necessary adjust the 95-kΩ resistor to obtain full-scale deflection.

Step 5. Measure the deflections produced for *each* value of R_x in Table 5-7. Make a neat sketch of your calibrated meter face.

Conclusion

What is the relationship between the value of R_x that produces a half-scale deflection (R_{mid}) and R, assuming the inequality $R \ll R_m + R_R$ is satisfied? What is the relationship between R_{mid} and the current supplied by the battery when the ohmmeter is zeroed? Why is this relationship important?

Table 5-7. Data for Experiment 5-2

Deflection (D)	Percent Deflection	R_x	Percent Deflection (Measured)
0	0	∞ Ω	
1/8	12.5		
1/4	25		
3/8	37.5		
1/2	50		
5/8	62.5		
3/4	75		
7/8	87.5		
1	100	0 Ω	

CHAPTER 6

Rectifier-Type AC Voltmeters

6-1 INTRODUCTION

A basic meter movement with appropriate resistors, batteries, and switches enables you to measure a wide range of dc current, dc voltage, and resistance. If a *low-frequency* ac current flows through a meter movement, the pointer will attempt to follow the current changes. Thus the pointer will swing in one direction and then the other. At *higher* frequencies the inertia of the coil assembly is large enough to prevent the pointer from following the current changes. In such a case the pointer deflects to the *average value* of the ac current. There are several types of meter movements that provide deflections proportional to *rms* rather than average values. Such meter movements are employed in "special" rather than general-purpose instruments. In the sections that follow you will learn how to design and calibrate elementary, general-purpose, ac voltmeters.

6-2 OBJECTIVES

At the end of this chapter you will be able to do the following:

- Explain what is meant by an average value.
- Explain what is meant by a transfer instrument.
- Design and calibrate half-wave, rectifier-type, ac voltmeters.
- Design and calibrate full-wave, rectifier-type, ac voltmeters.
- Understand the loading effects of rectifier-type ac voltmeters.

- Understand how the nonlinear properties of real diodes affect the measurement of small ac signals.

6-3 AVERAGE VALUES

Suppose you obtained a part-time job. During the first week you earned $200.00, and during the second week $100.00. What is your average salary for the two-week period? The correct answer, of course, is $150.00/week. Most people would have little trouble arriving at this figure due to the fact virtually everyone has acquired a basic "understanding" of money. We will begin our discussion of average values by graphically representing the previous question in Fig. 6-1. Given a waveform, how do you specify its average value? With reference to Fig. 6-1 a little thought will reveal the fact that

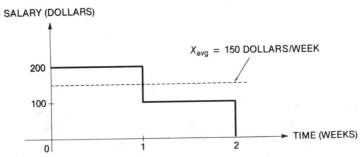

Fig. 6-1. Average value.

the average value is simply the *area* under the waveform divided by the *base*. Thus, in Fig. 6-1,

$$\text{average salary} = \frac{(200\ \text{dollars})(1\ \text{week}) + (100\ \text{dollars})(1\ \text{week})}{2\ \text{weeks}}$$

$$= 150\ \text{dollars/week}$$

In general, for any waveform

$$X_{avg} = \frac{A}{b} \tag{6-1}$$

where
X_{avg} = average value,
A = area,
b = base.

If the waveform's geometry is simple, then it is easy to compute the average value.

EXAMPLE 6-1

Determine the average value of the current in Fig. 6-2.

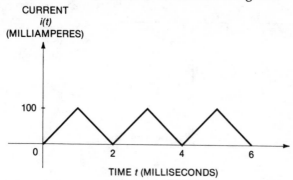

Fig. 6-2. Waveform for Example 6-1.

For a periodic waveform the base b is taken as the period (T). Thus

$$T = b = 2\,\text{ms}$$

From "advanced" mathematics courses you know that the area under a triangle is ½bh. Therefore

$$A = \tfrac{1}{2}\,bh$$
$$A = \tfrac{1}{2}\,(2\,\text{ms})(100\,\text{mA})$$
$$A = 100\,\text{ms} \cdot \text{mA}$$

Thus the *average* current (I_{avg}) is

$$I_{\text{avg}} = \frac{A}{b} = \frac{100\,\text{ms} \cdot \text{mA}}{2\,\text{ms}} = 50\,\text{mA}$$

EXAMPLE 6-2

Determine the average value of the voltage in Fig. 6-3. Here $T = b = 4\,\mu\text{s}$ and

$$A = (20\,\text{V})(2\,\mu\text{s}) + (-20\,\text{V})(2\,\mu\text{s})$$
$$= 40\,\text{V} \cdot \mu\text{s} - 40\,\text{V} \cdot \mu\text{s}$$
$$= 0\,\text{V} \cdot \mu\text{s}$$

Therefore

$$V_{\text{avg}} = \frac{A}{b} = \frac{0\,\text{V} \cdot \mu\text{s}}{4\,\mu\text{s}} = 0\,\text{V}$$

Since the voltage in Fig. 6-3 is centered around zero, the area above zero (positive area) equals the area below zero (negative area). Thus the *net area* averaged over one period *is zero*. From Equation 6-1 you can see that a net area of zero means the average value must also be zero. Remember, most meter movements deflect to the average value of the current through them.

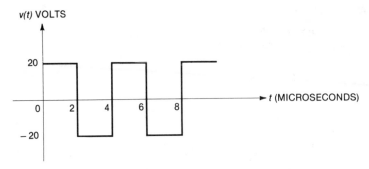

Fig. 6-3. Waveform for Example 6-2.

EXAMPLE 6-3

A 100-mA meter movement is used to measure the current in Fig. 6-2. What will the meter movement read?

The average current was calculated to be 50 mA in Example 6-1. Since the meter movement deflects to the average current it will read 50 mA.

EXAMPLE 6-4

A 0- to 50-V dc voltmeter is used to measure the voltage in Fig. 6-3. What will the voltmeter read?

The amount of deflection is proportional to the average current through the voltmeter's meter movement. Since the average current is zero, the average voltage across the voltmeter (V_{avg}) must also be zero. Thus the voltmeter reads 0 V.

Another name for *average* value is *dc* value. This is the case because *a dc instrument deflects to the average value of the quantity being measured.*

6-4 TRUE-RMS MOVEMENTS

A true-rms movement provides a deflection proportional to *current squared*. Such a movement is easily calibrated to read the rms value of the current (or voltage) being measured. Recall from basic electricity, or the review of ac electricity provided in Appendix D, that the rms value of a current is defined as the square root of the average of the squares of the instantaneous values taken over one period. The *rms value* of a current (or voltage) *represents the equivalent amount of dc with respect to the heating effect produced.*

The *iron-vane* and *electrodynamometer* movements discussed in Chapter 2 are examples of current-squared devices. A third type of current-squared device is the *thermocouple ammeter* illustrated in Fig. 6-4. Notice in Fig. 6-4 that a thermocouple consists of two conductors made from *different* metals which are joined at one end.

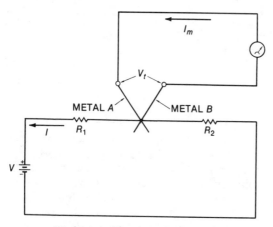

Fig. 6-4. A thermocouple ammeter.

Whenever two dissimilar metals are joined, a voltage is developed between the opposite ends of the conductors (Seebeck effect). The magnitude of the voltage depends on the metals used and the junction temperature. If the temperature of the junction increases, the voltage generated by the thermocouple will also increase.

In Fig. 6-4 the current (I) in the circuit heats the junction of the thermocouple. As a result the voltage V_t is generated by the thermocouple. The magnitude of V_t and the resistance of the ammeter determine the value of I_m through the meter movement, since $I_m = V_t/R_m$. Thus the ammeter can easily be calibrated to read the rms value of I. A significant advantage of true-rms instruments is that they can be used to measure *nonsinusoidal* waveforms just as easily as dc or sinusoidal waveforms. As you will see later, rectifier and peak-type ac voltmeters are calibrated based on the assumption that the input voltage is a sine wave. If you attempt to measure nonsinusoidal voltages with a rectifier or peak-type ac voltmeter, it is unlikely that the reading obtained will be accurate. This is *not* the case with true-rms voltmeters.

Another advantage of true-rms instruments is that they can be used as *transfer* instruments. The electrodynamometer movement for example is often calibrated on dc, and then employed directly for ac measurements.

6-5 RECTIFIER-TYPE AC VOLTMETERS (HALF-WAVE)

An elementary half-wave, rectifier-type, ac voltmeter is illustrated in Fig. 6-5. Notice that the instrument consists of the series combination of a multiplier resistor (R_s), diode (D), and meter movement.

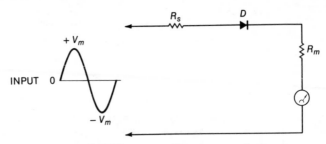

Fig. 6-5. Half-wave, rectifier-type, ac voltmeter.

From your study of electronic devices you know that a diode acts like an electronic switch. Fig. 6-6 summarizes the properties of an *ideal* diode. If an ideal diode is *forward biased* (anode more positive than the cathode) it acts as a short circuit ($r_D = 0$). This is illustrated in Fig. 6-6B. Similarly, when an ideal diode is *reverse biased* (anode more negative than the cathode) it acts as an open circuit ($r_D = \infty$), which is illustrated in Fig. 6-6C. The I/V curve for an ideal diode is illustrated in Fig. 6-6D. A typical I/V curve for a *real* diode is illustrated in Fig. 6-7. An insight into the significant differences between an ideal diode and a real diode can be obtained by comparing Fig. 6-7 with Fig. 6-6. After a reasonable amount of time has elapsed for contemplation we conclude that real diodes differ from ideal diodes in the following respects:

1. For small forward voltages ($V \leqq V_K$) the current through a real diode (I_F) is negligible. Thus, for a real diode to conduct, the voltage across it must exceed V_K, where V_K is called the *knee* voltage. For germanium (Ge) diodes $V_K \cong 0.3$ V, and for silicon (Si) diodes $V_K \cong 0.7$ V.

2. When a real diode does conduct ($V > V_K$) it exhibits a finite forward resistance (r_F). Since the diode curve above the knee is reasonably linear you can estimate r_F from the diode's I/V curve as follows:

$$r_F = \frac{\Delta V}{\Delta I} \qquad (6\text{-}2)$$

where ΔI is the very small change in current produced by a very small change of voltage ΔV.

3. When a real diode is reverse biased a small reverse current (I_R) flows. This suggests that a real diode acts as a *large resistance* (R_r) when reverse biased rather than an infinite resistance (open circuit).

In instruments real diodes will closely approximate the properties of ideal diodes if the following conditions are met:

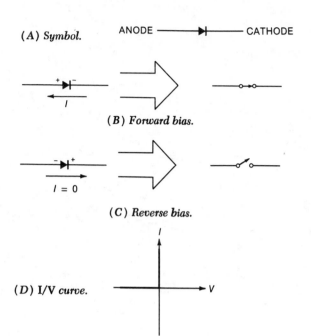

(A) Symbol.   ANODE ——————▶|—————— CATHODE

(B) Forward bias.

(C) Reverse bias.

(D) I/V curve.

Fig. 6-6. Properties of an ideal diode.

$$V \gg V_K \qquad (6\text{-}3)$$
$$R_f \gg r_F \qquad (6\text{-}4)$$
$$R_R \gg R_r \qquad (6\text{-}5)$$

where

V = voltage driving the forward-biased diode,
V_K = diode knee voltage,
R_f = equivalent resistance in series with the forward-biased diode,
r_F = forward resistance of the diode,
R_R = reverse resistance of the diode,
R_r = equivalent resistance in series with the reverse-biased diode.

How does the ac voltmeter in Fig. 6-5 work? During the positive half-cycle of the input voltage, diode D is forward biased. Thus the diode will conduct. This is illustrated in Fig. 6-8A. During the negative half-cycle diode D is reverse biased. In this case the diode will approximate an open circuit. This is illustrated in Fig. 6-8B. Since the diode can only conduct in one direction the current in the ac voltmeter in Fig. 6-5 will be pulsating dc characteristic of a half-wave rectifier. This is illustrated in Fig. 6-9. The meter movement in Fig. 6-5 will deflect to the *average value* of the current illustrated in Fig. 6-9. Recall that average value is simply area divided by base.

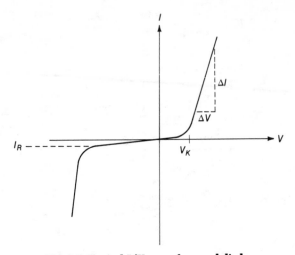

Fig. 6-7. Typical I/V curve for a real diode.

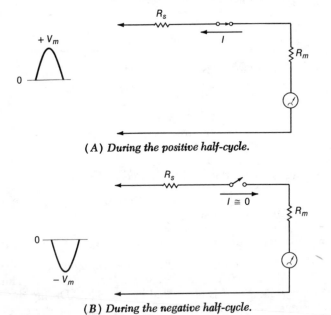

(A) During the positive half-cycle.

(B) During the negative half-cycle.

Fig. 6-8. Equivalent circuit for the voltmeter in Fig. 6-5.

Since the geometry of the current in Fig. 6-9 is more complicated than the waveforms encountered previously we will *not* derive the average value. The results of this derivation are quite simple, however. As you can see in Fig. 6-9 the average (dc) value for a half-

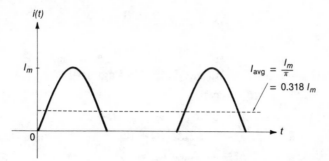

Fig. 6-9. Current in a half-wave, rectifier-type, ac voltmeter (sinusoidal input).

wave signal is simply the maximum value multiplied by 0.318. Thus

$$I_{\text{avg}} = 0.318I_m \tag{6-6}$$

Similarly, if the waveform in Fig. 6-9 were a voltage

$$V_{\text{avg}} = 0.318V_m \tag{6-7}$$

Since the deflection is proportional to average current (I_{avg}), and since the average current depends on the amplitude of the ac input voltage, it is possible to calibrate the meter face to read voltage directly. Normally the meter face is calibrated to indicate the *rms* value of the input voltage. In order to illustrate how this is done we will derive some very useful relationships.

Neglecting the resistance of the diode in Fig. 6-5, the input resistance of the voltmeter (R_{in}) is, by inspection,

$$R_{\text{in}} = R_s + R_m$$

Therefore

$$R_s = R_{\text{in}} - R_m \tag{6-8}$$

Maximum diode current occurs when the diode is forward biased and the input voltage is maximum. Thus

$$I_m = \frac{V_m}{R_s + R_m}$$

From Equation 6-6 the *average* (dc) diode current is

$$I_{\text{avg}} = 0.318I_m$$

$$I_{\text{avg}} = \frac{0.318V_m}{R_s + R_m}$$

Solving for V_m yields

$$I_{\text{avg}} (R_s + R_m) = 0.318V_m$$

$$V_m = \frac{I_{\text{avg}}(R_s + R_m)}{0.318}$$

$$V_m = 3.14 I_{\text{avg}}(R_s + R_m)$$

The rms value for the *sinusoidal input* voltage is

$$\begin{aligned} V_{\text{rms}} &= 0.707 V_m \\ &= 0.707(3.14) I_{\text{avg}}(R_s + R_m) \\ &= 2.22 I_{\text{avg}}(R_s + R_m) \\ &= 2.22 I_{\text{avg}} R_{\text{in}} \end{aligned}$$

We wish to use the voltmeter in Fig. 6-5 to measure *sinusoidal* voltages up to some maximum rms value (V_{FS}). Thus, in order for the meter movement to deflect full-scale when V_{FS} is applied to the input of the voltmeter, the average current (I_{avg}) must equal the dc current required for full-scale deflection (I_{FS}). Substituting V_{FS} for V_{rms} and I_{FS} for I_{avg} yields

$$V_{FS} = 2.22 I_{FS}(R_s + R_m) \qquad (6\text{-}9)$$

Solving Equation 6-9 for $R_s + R_m$ yields

$$R_s + R_m = \frac{V_{FS}}{2.22 I_{FS}} = R_{\text{in}}$$

In order to keep our results similar to those provided for dc voltmeters we will define the sensitivity of a half-wave (hw), rectifier-type, ac voltmeter as follows:

$$S_{hw} = \frac{1}{2.22 I_{FS}} = \frac{0.45}{I_{FS}} \qquad (6\text{-}10)$$

Thus

$$R_s + R_m = R_{\text{in}} = \frac{1}{2.22 I_{FS}} \times V_{FS}$$

so that

$$R_{\text{in}} = S_{hw} V_{FS} \qquad (6\text{-}11)$$

The utility of the above relationships is illustrated in Example 6-5.

EXAMPLE 6-5

Assuming a 50-μA 2-kΩ movement is available, design:

   (a) A 0- to 10-V dc voltmeter.
   (b) A 0- to 10-V-rms half-wave, rectifier-type, ac voltmeter.

   (a)

$$S = \frac{1}{I_{FS}} = \frac{1}{50\ \mu A} = 20\ \frac{k\Omega}{V}$$

$$R_{in} = SV_{FS}$$

$$= \left(20\frac{k\Omega}{V} \right)(10 \text{ V}) = 200 \text{ k}\Omega$$

$$R_s = R_{in} - R_m$$

$$= 200 \text{ k}\Omega - 2 \text{ k}\Omega = 198 \text{ k}\Omega$$

(b)

$$S_{hw} = \frac{0.45}{I_{FS}} = \frac{0.45}{50 \ \mu\text{A}} = 9 \ \frac{k\Omega}{V}$$

$$R_{in} = S_{hw}V_{FS}$$

$$= \left(9 \ \frac{k\Omega}{V} \right)(10 \text{ V})$$

$$= 90 \text{ k}\Omega$$

$$R_s = R_{in} - R_m$$

$$= 90 \text{ k}\Omega - 2 \text{ k}\Omega$$

$$= 88 \text{ k}\Omega$$

Both the dc and ac voltmeters can easily be packaged into a single instrument as illustrated in Fig. 6-10.

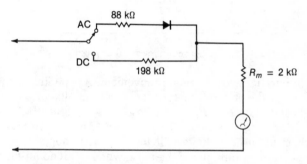

Fig. 6-10. A dc, half-wave, ac-type voltmeter for Example 6-5.

Equation 6-9 indicates that average current is directly proportional to the rms value of the ac input voltage. This situation is similar to that of a dc voltmeter in which dc current is directly proportional to the dc input voltage. Thus, if the diode in Fig. 6-10 is ideal, or closely approximates an ideal diode, then the *same scale* could be used for dc and ac voltage measurements. Many inexpensive commercial instruments do in fact employ the same scale for dc and ac voltage measurements.

In our derivation of Equation 6-9, however, we neglected the non-ideal properties of the diode. Recall that one of the conditions which must be satisfied in order for a real diode to closely approximate an ideal diode is that the voltage driving the diode (V) must be large compared to the diode's knee voltage (V_K). When you attempt to measure small ac voltages with the voltmeter in Fig. 6-10 this condition is *not* satisfied. From the typical diode curve in Fig. 6-7 you can see that a small ac voltage establishes the diode's operating point near or below the knee. Operation in this *nonlinear* region of the curve means that average current is *not* directly proportional to the rms value of the ac input voltage. Thus the deflection which results is very small and does not accurately reflect the amplitude of the ac input voltage in a linear manner. The meter reads *less* than it should. For this reason most commercial ac voltmeters do *not* provide ac voltage ranges less than 0 to 1.5 V.

A partial solution to this problem is illustrated by the circuit in Fig. 6-11. Notice in Fig. 6-11 that a shunt resistance (R_{sh}) has been

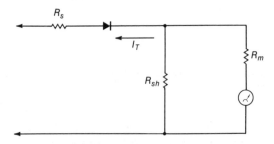

Fig. 6-11. Compensating for small ac input voltages.

placed in parallel with the meter movement. As you know, the series equivalent of the shunted meter movement acts like a meter movement whose $R_m' = R_{sh} \parallel R_m$, and whose $I_{FS} = I_T$. Since $R_m' < R_m$, the total resistance in the circuit is decreased. Thus the diode current for a given ac input voltage will *increase*, it is hoped, moving the diode's operating point above the knee, when small ac input voltages are applied to the input of the voltmeter.

Since the sensitivity of a half-wave, rectifier-type, ac voltmeter is only 45 percent of the dc sensitivity, the ac input resistance will only be 45 percent of the dc input resistance for the same range. This means that loading effects will be more pronounced when you measure ac voltages than dc voltages. Remember, when a meter movement is shunted, the sensitivity of the series equivalent is less than the original unshunted movement. Thus the disadvantage of the circuit in Fig. 6-11 is that sensitivity has been decreased. This is the price you must pay in order to compensate for the nonlinear properties of real diodes.

A practical version of a half-wave, rectifier-type, ac voltmeter typical of those found in commercial instruments is illustrated in

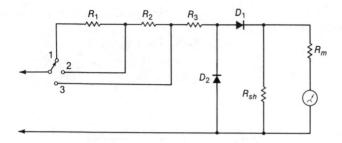

Fig. 6-12. Practical half-wave, rectifier-type, ac voltmeter.

Fig. 6-12. Resistors R_1, R_2, and R_3 form a series chain of multiplier resistances for the three ranges 1, 2, and 3. Resistor R_{sh} is employed to compensate for the nonlinear properties of D_1 as discussed previously. Recall that when a real diode is reverse biased a small reverse current (I_R) flows through it. If diode D_2 were not in the circuit, a small reverse current would flow through D_1 on the negative half-cycle of the ac input voltage. Thus the purpose of diode D_2 is to prevent this small reverse current from flowing through diode D_1 (and the meter movement) on the negative half-cycle of the ac input voltage. During the negative half-cycle diode D_2 is forward biased, thus shunting current around diode D_1 and the meter movement.

EXAMPLE 6-6

Design an ac voltmeter like the one shown in Fig. 6-12. A 50-μA, 2-kΩ meter movement is available for this modest task. The voltmeter should provide the following ranges: 0–5 V, 0–10 V, and 0–50 V rms.

A reasonable compromise between the decreased sensitivity resulting from shunting a meter movement and employing the shunt to compensate for diode D_1's nonlinear characteristics is to *make the shunt resistance equal to the resistance of the unshunted meter movement.* Thus

$$R_{sh} = R_m = 2 \text{ k}\Omega$$

The shunted meter movement has a series equivalent whose $R_m' = R_{sh} \parallel R_m$ and whose $I_{FS} = I_T$. Thus

$$R_m' = 2 \text{ k}\Omega \parallel 2 \text{ k}\Omega = 1 \text{ k}\Omega$$

Since $R_m = R_{sh}$, the total current (I_T) required for full-scale deflection of the shunted movement is

$$I_T = 100 \text{ μA} = I_{FS} \qquad \text{(shunted movement)}$$

The ac sensitivity for the voltmeter is

$$S_{hw} = \frac{0.45}{I_{FS}} = 4.5 \text{ } \frac{\text{k}\Omega}{\text{V}}$$

149

Knowing the sensitivity, we can easily calculate the input resistance required for each range.

The 0- to 5-V range (switch position 3, Fig. 6-13):

$$R_{in} = S_{hw}V_{FS}$$
$$= \left(4.5\frac{k\Omega}{V}\right)(5\text{ V})$$
$$= 22.5\text{ k}\Omega$$

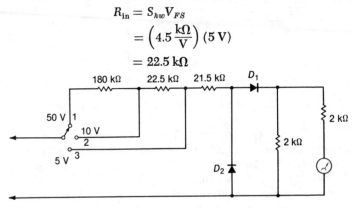

Fig. 6-13. The ac voltmeter for Example 6-6.

The 0- to 10-V range (switch position 2, Fig. 6-13):

$$R_{in} = S_{hw}V_{FS}$$
$$= \left(4.5\frac{k\Omega}{V}\right)(10\text{ V})$$
$$= 45\text{ k}\Omega$$

The 0- to 50-V range (switch position 3, Fig. 6-13):

$$R_{in} = S_{hw}V_{FS}$$
$$= \left(4.5\frac{k\Omega}{V}\right)(50\text{ V})$$
$$= 225\text{ k}\Omega$$

With reference to Fig. 6-13,

$$R_3 = 22.5\text{ k}\Omega - 1\text{ k}\Omega = 21.5\text{ k}\Omega$$
$$R_2 = 45\text{ k}\Omega - (R_3 + R_m')$$
$$= 45\text{ k}\Omega - 22.5\text{ k}\Omega$$
$$= 22.5\text{ k}\Omega$$
$$R_1 = 225\text{ k}\Omega - (R_2 + R_3 + R_m')$$
$$= 225\text{ k}\Omega - 45\text{ k}\Omega$$
$$= 180\text{ k}\Omega$$

The final design is presented in Fig. 6-13.

Many commercial instruments employ ac voltmeters similar to the voltmeter illustrated in Fig. 6-12. Higher-quality instruments will often provide a *separate* scale for ac voltage measurements. Recall that the purpose of the shunt resistance in Fig. 6-12 is to compensate for the nonlinear properties of diode D_1. This compensation is *not* perfect—especially for very small ac input voltages, where average current is *not* directly proportional to the rms value of the ac input voltage. Thus, *the separate ac voltage scale is a nonlinear scale.* The greatest deviation from the dc voltage scale occurs for very small ac input voltages.

EXAMPLE 6-7

The voltmeter in Fig. 6-10 is employed to measure the voltage between A and B in Fig. 6-14. Determine:

(*a*) The original voltage.
(*b*) The percent accuracy and loading error.
(*c*) The measured voltage.

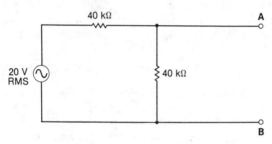

Fig. 6-14. Circuit for Example 6-7.

(*a*) In Fig. 6-14 the original voltage will equal the Thevenin equivalent voltage. Thus

$$V_o = V_{TH} = 10 \text{ V rms}$$

(*b*) Fig. 6-15 illustrates the Thevenin equivalent circuit of Fig. 6-14 with the load (voltmeter) connected. Note that the input resistance (R_{in}) is only 90 kΩ for the ac range. Thus the accuracy is

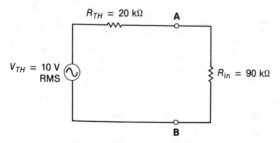

Fig. 6-15. Thevenin equivalent circuit for Example 6-7 with voltmeter (load) connected.

$$a = \frac{R_{in}}{R_{in} + R_{TH}} = \frac{90 \text{ k}\Omega}{90 \text{ k}\Omega + 20 \text{ k}\Omega} = 0.818$$
$$= 81.8\%$$

$$e = 1 - a = 1 - 0.818 = 0.181 = 18.1\%$$

(c)

$$V_m = aV_o$$
$$= 0.818(10)$$
$$= 8.18 \text{ V rms}$$

Example 6-7 illustrates the fact that you can determine the loading effects of ac voltmeters in a manner similar to that of dc voltmeters. The essential difference is, of course, that the loading effects of ac voltmeters are more pronounced. Recall that the input resistance of a half-wave, rectifier-type, ac voltmeter is only 45 percent of the dc input resistance for the same range. This situation can be improved considerably by utilizing a full-wave rectifier instead of a half-wave rectifier. This modification is the topic of the next section.

6-6 RECTIFIER-TYPE AC VOLTMETERS (FULL-WAVE)

A full-wave, rectifier-type, ac voltmeter is illustrated in Fig. 6-16. This circuit employs a full-wave *bridge* rectifier, which is popular in measuring instruments. During the positive half-cycle, diodes D_1 and D_3 are forward biased while diodes D_2 and D_4 are reverse biased. Assuming ideal diodes we can picture the equivalent circuit shown in Fig. 6-17A. The arrows in Fig. 6-17 indicate the path of

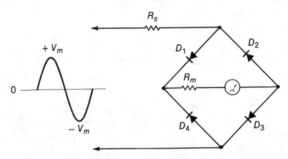

Fig. 6-16. Full-wave, rectifier-type, ac voltmeter.

electron flow. During the negative half-cycle diodes D_2 and D_4 are forward biased, and diodes D_1 and D_3 are reverse biased. This is illustrated in Fig. 6-17B. Notice in Fig. 6-17 that the direction of current through the meter movement is the *same* for *each* half-cycle. Thus the current through the meter movement in Fig. 6-16 will be

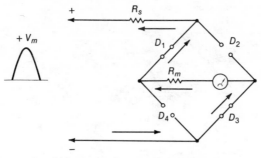

(A) During the positive half-cycle.

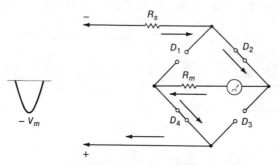

(B) During the negative half-cycle.

Fig. 6-17. Equivalent circuit for the voltmeter in Fig. 6-16.

pulsating dc characteristic of a full-wave rectifier. This is illustrated in Fig. 6-18. Naturally, the meter movement in Fig. 6-16 will deflect to the average value of the current illustrated in Fig. 6-18. If you compare Figs. 6-18 and 6-19, it should be obvious that the average value of a full-wave signal is twice that of a half-wave signal. Thus

$$I_{avg} = 0.636I_m \qquad (6\text{-}12)$$

and

$$V_{avg} = 0.636V_m \qquad (6\text{-}13)$$

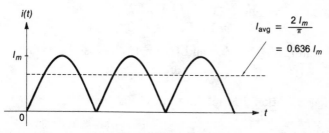

Fig. 6-18. Current in a full-wave, rectifier-type, ac voltmeter (sinusoidal input).

153

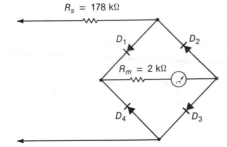

Fig. 6-19. The ac voltmeter for Example 6-8.

In a manner directly analogous to the derivation provided for the half-wave, rectifier-type, ac voltmeter we will quickly (but efficiently) derive the appropriate relationships for the full-wave, rectifier-type, ac voltmeter:

$$R_{\text{in}} = R_s + R_m \quad \text{or} \quad R_s = R_{\text{in}} - R_m$$

Also,

$$I_m = \frac{V_m}{R_s + R_m}$$

and

$$I_{\text{avg}} = 0.636 I_m$$
$$= \frac{0.636 V_m}{R_s + R_m}$$

Solving for V_m yields

$$I_{\text{avg}}(R_s + R_m) = 0.636 V_m$$
$$V_m = \frac{I_{\text{avg}}(R_s + R_m)}{0.636}$$
$$= 1.57 I_{\text{avg}}(R_s + R_m)$$

Since $V_{\text{rms}} = 0.707 V_m$,

$$V_{\text{rms}} = 0.707 (1.57) I_{\text{avg}}(R_s + R_m)$$
$$= 1.11 I_{\text{avg}}(R_s + R_m)$$

Substituting V_{FS} for V_{rms} and I_{FS} for I_{avg} yields

$$V_{FS} = 1.11 I_{FS}(R_s + R_m) \tag{6-14}$$

Solving Equation 6-14 for $R_s + R_m$ yields

$$R_s + R_m = \frac{V_{FS}}{1.11 I_{FS}} = R_{\text{in}}$$

We define the sensitivity of a full-wave (fw), rectifier-type, ac voltmeter to be

$$S_{fw} = \frac{1}{1.11 I_{FS}} = \frac{0.9}{I_{FS}} \qquad (6\text{-}15)$$

Thus

$$R_{\text{in}} = R_s + R_m = \frac{1}{1.11 I_{FS}} \times V_{FS}$$

$$R_{\text{in}} = S_{fw} V_{FS} \qquad (6\text{-}16)$$

Notice that the sensitivity of a full-wave, rectifier-type, ac voltmeter is *double* the sensitivity of a half-wave, rectifier-type, ac voltmeter.

EXAMPLE 6-8

Design a 0- to 10-V-rms full-wave, rectifier-type, ac voltmeter. Assume a 50-μA 2-kΩ meter movement is available for the design. If this voltmeter is used to measure the voltage between A and B in Fig. 6-14, what will it read?

$$S_{fw} = \frac{0.9}{I_{FS}} = \frac{0.9}{50\ \mu A} = 18\frac{k\Omega}{V}$$

$$R_{\text{in}} = S_{fw} V_{FS}$$

$$= \left(18\frac{k\Omega}{V} \right)(10\ V) = 180\ k\Omega$$

$$R_s = R_{\text{in}} - R_m$$
$$= 180\ k\Omega - 2\ k\Omega$$
$$= 178\ k\Omega$$

From Example 6-7, $R_{TH} = 20$ kΩ and $V_{TH} = V_o$ 10 V rms. Therefore

$$a = \frac{R_{\text{in}}}{R_{\text{in}} + R_{TH}} = \frac{180\ k\Omega}{180\ k\Omega + 20\ k\Omega} = 0.9 = 90\%$$

$$V_m = aV_o$$
$$= 0.9(10\ V)$$
$$= 9\ V\ rms$$

The design is illustrated in Fig. 6-19. From Equations 6-14, 6-15, and 6-16 you can see that the sensitivity and input resistance of a full-wave, rectifier-type, ac voltmeter is 90 percent of the dc sensitivity and input resistance for the same range. Thus the loading effects of a full-wave, rectifier-type, ac voltmeter are only *slightly* more pronounced than those of a dc voltmeter. Note that *two diodes* are forward biased on alternate half-cycles in a bridge rectifier as opposed to *one* diode in the half-wave version. Thus the problem encountered due to the nonlinear properties of a diode in the half-wave, rectifier-type, ac voltmeter is more critical in the full-wave, rectifier-type, ac voltmeter. Recall that this problem is significant only when you attempt to measure *small* ac voltages.

6-7 REVIEW OF OBJECTIVES

Most meter movements provide deflections proportional to *average* current. In rectifier-type voltmeters a half- or full-wave rectifier is employed to obtain a pulsating dc current. The pulsating dc current flows through a meter movement which deflects to the average value of the current. Normally, rectifier-type instruments are calibrated to indicate the rms value of *sinusoidal* input voltages. Loading effects of rectifier-type ac voltmeters are similar to those of dc voltmeters. The essential difference is that loading effects are more serious for ac measurements than dc measurements due to the fact that the ac sensitivity of rectifier-type voltmeters is less than the dc sensitivity.

The iron-vane-electrodynamometer and thermocouple movements provide deflections proportional to *current squared*. Thus these movements can be employed in transfer and true-rms instruments. Due to the fact that real diodes require a forward voltage equal to or greater than the knee voltage to become forward biased it is difficult to measure small ac voltages.

6-8 QUESTIONS

1. What is a transfer instrument?
2. Why is an average value of a current or voltage called the dc value?
3. What is the average value of a 200-V-peak sine or cosine wave?
4. Where could you use a thermocouple ammeter?
5. What is an advantage of a full-wave, rectifier-type voltmeter over a half-wave, rectifier-type voltmeter? A disadvantage?
6. What are typical knee voltages for germanium and silicon diodes?
7. What conditions must be met for real diodes to approximate ideal diodes?
8. Describe the purpose of D_2 in Fig. 6-12. Are there any disadvantages in including D_2 in the circuit?
9. When Dr. Courtine was employed by the U.R. Ripoffski power utility, he suggested that customers be billed for the average electricity used. Why do you think his suggestion was *not* adopted?

6-9 PROBLEMS

1. What is the input resistance of a dc voltmeter, a half-wave, rectifier-type, ac voltmeter, and a full-wave, rectifier-type, ac voltmeter for the following ranges: 0–10 V, 0–50 V, and 0–100 V, assuming a 1-mA, 50-Ω movement is employed in each instrument?
2. What will a 50-μA, 1-kΩ meter movement read if it is used to measure the current illustrated in Fig. 6-20A? In Fig. 6-20B?

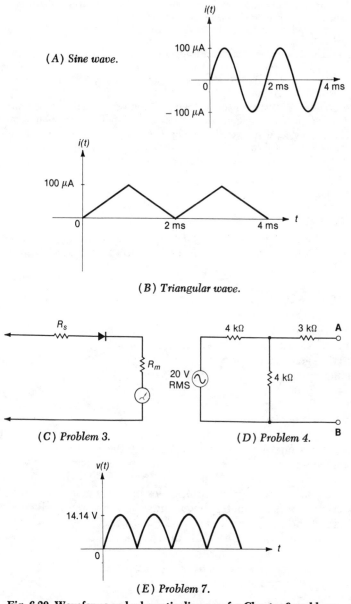

(A) Sine wave.

(B) Triangular wave.

(C) Problem 3.

(D) Problem 4.

(E) Problem 7.

Fig. 6-20. Waveforms and schematic diagrams for Chapter 6 problems.

3. The meter movement in Fig. 6-20C has an R_m of 500 Ω and an I_{FS} of 200 μA. What is the value of R_s if sinusoidal voltages up to 20 V rms are to be measured?

4. Assume that the voltmeter in Problem 3 is used to measure the voltage between A and B in Fig. 6-20D. Determine:
   (a) The voltage between A and B before the voltmeter is inserted in the circuit.
   (b) The percent accuracy and loading error.
5. Design a voltmeter similar to that of Fig. 6-12 that has the following ranges: 0–5 V, 0–20 V, and 0–40 V. Assume that a 100-μA, 2-kΩ meter movement is used in the design.
6. A 150-μA, 1-kΩ movement is used in a full-wave, rectifier-type voltmeter. What value of R_s is required if the maximum input voltage is 50 V rms?
7. The rms value of the voltage in Fig. 6-20E is $V_m/\sqrt{2}$. If the voltmeter in Problem 3 is used to measure this voltage, will the reading obtained be accurate? Why? What will the voltmeter read?

6-10 EXPERIMENT 6-1

Objective

The objective of this experiment is to determine the frequency above which a meter movement deflects to the *average value* of the signal being measured.

Material Required

50-μA meter movement
Fixed resistor or resistor decade—you will calculate the value required as part of the experiment
Function generator
Oscilloscope

Introduction

For frequencies higher than a minimum frequency, the pointer of most meter movements will provide a deflection proportional to average current. This is the case because the inertia of the coil assembly is large enough to prevent the pointer from following the variations of the current. The specific frequency at which this occurs depends on the mechanical properties of the meter movement you are using. For many meter movements the minimum frequency will lie within the range of 10 Hz to 30 Hz. In this experiment you will design an elementary dc voltmeter which will be employed to determine the minimum frequency. Recall that the average current through the meter movement will be zero if the average voltage across the dc voltmeter is zero.

Procedure

Step 1. Calculate the value of R_s in Fig. 6-21 so that full-scale deflection occurs if the input voltage is 10 Vdc.
   $R_s =$ _____

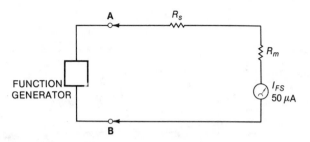

Fig. 6-21. Schematic diagram for Experiment 6-1.

Step 2. Using an oscilloscope, adjust the function generator for a sinusoidal output of 5 V peak at 100 Hz.

Step 3. Measure the output voltage of the function generator with the dc voltmeter.
$V_{\text{voltmeter}} =$ _____

Step 4. Reduce the frequency until the pointer *starts* oscillating.

Step 5. Gradually increase the frequency until the pointer stops oscillating—this is $f_{\min}$. Record the value of $f_{\min}$.
$f_{\min} =$ _____

Step 6. If your function generator has a dc offset, adjust the output as follows:
$f = 1 \text{ kHz}$
offset $= +5 \text{ Vdc}$
$v_0 = 5 \text{ V peak}$

Step 7. Measure the output of the function generator with the dc voltmeter.
$V_{\text{voltmeter}} =$ _____

Discussion

The average value of a sinusoid is zero. Since the average voltage across the dc voltmeter in Step 3 was zero, the average current through the meter movement was also zero. When $f \gtrsim f_{\min}$ the deflection of most meter movements is proportional to average current. Thus, when you measured the output voltage of the function generator at 100 Hz in Step 3 you should *not* have observed any deflection since 100 Hz $> f_{\min}$. If R_s was shorted (0 Ω) in Fig. 6-21 and the amplitude of the input voltage was increased substantially, the average current through the meter movement would *still be zero*. In this case, however, the *peak* current through the meter movement could be large enough to damage the meter movement! This is an "interesting" situation where an instrument correctly reads zero, but is damaged or destroyed in the process. *Such "interesting" situations should be avoided in practice!*

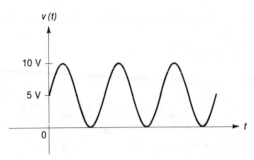

Fig. 6-22. Waveform discussed in Experiment 6-1.

If your function generator had a dc offset you should have observed that the dc voltmeter deflected to the offset voltage, since this *was* the average value. To understand why this is so, consider Fig. 6-22, which illustrates a 5-V-peak sinusoid offset by +5 Vdc. Notice that the sinusoid "rides on" the 5-Vdc level. Since the average value of the sinusoid is zero, it follows that the average value of the offset waveform *is* the dc offset level, which is 5 V in this example.

Conclusion

Due to the nature of this experiment our conclusion was presented in the Discussion section. Thus you should reread that section in order to see what kinds of conclusions you might write. In addition, you should indicate whether or not your measurements agree with the predictions made in the Discussion section. If they do not, attempt to explain why this is the case.

6-12 EXPERIMENT 6-2

Objective

The objective of this experiment is to design a half-wave, rectifier-type, ac voltmeter. In addition, the problems encountered due to the nonideal properties of real diodes will be explored.

Material Required

50-μA meter movement
Resistor decades: 100-kΩ, 10-kΩ, 1-kΩ, and 100-Ω steps
Function generator
Oscilloscope
1N4001 diode (almost any general-purpose rectifier diode will do)

Introduction

The average current in a half-wave rectifier is proportional to the amplitude of the input voltage. Since deflection is proportional to the average current through a meter movement, it is possible to calibrate the meter movement in a half-wave, rectifier-type voltmeter to read peak, peak-to-peak, or rms values of the input voltage. Normally, rectifier-type voltmeters are calibrated to indicate the rms values of sinusoidal input voltages. The appropriate equations for a half-wave, rectifier-type voltmeter derived in the text include the following:

$$V_{FS} = 2.22I_{FS}(R_s + R_m) \qquad (6\text{-}9)$$

$$S_{hw} = \frac{0.45}{I_{FS}} \qquad (6\text{-}10)$$

$$R_{in} = S_{hw}V_{FS} \qquad (6\text{-}11)$$

$$R_s = R_{in} - R_m$$

The derivation of the above equations *assumed* an ideal diode. Recall that for a real diode to approximate an ideal diode, the following relationships must be true:

$$V \gg V_K \qquad (6\text{-}3)$$

$$R_f \gg r_F \qquad (6\text{-}4)$$

$$R_R \gg R_r \qquad (6\text{-}5)$$

where
V = voltage driving the forward-biased diode,
V_K = diode knee voltage,
R_f = equivalent resistance in series with forward-biased diode,
r_F = forward resistance of the diode,
R_R = reverse resistance of the diode,
R_r = equivalent resistance in series with reverse-biased diode.

In this experiment, you will design a 0- to 7.07-V-rms, half-wave, rectifier-type, voltmeter. Fig. 6-23 illustrates how the meter face could be calibrated. Notice that the rms values for the third scale in Fig. 6-23 are 0.707 times the corresponding peak values provided on the second scale. The first scale simply indicates how the 50-μA movement is marked off in uniform divisions. We have included a scale to indicate peak values, because you will use an oscilloscope in this experiment to determine the degree of agreement between actual deflections and the deflections predicted by the calibrated scales in Fig. 6-23. Recall that it is much easier to read peak values with an oscilloscope than rms values. Table 6-1 summarizes the deflections predicted by the calibrated scales in Fig. 6-23.

Procedure

Step 1. Using Equations 6-10 and 6-11 calculate the sensitivity and

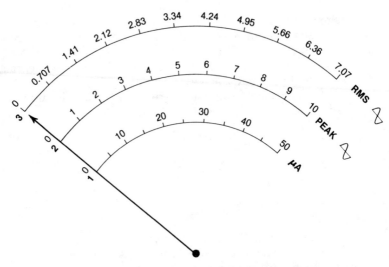

Fig. 6-23. Calibrated meter face for Experiment 6-2.

Table 6-1. Expected Deflections for Experiment 6-2

V_{rms} (Volts)	V_m (Volts)	D_{scale} (Percent)
7.07	10	100
6.36	9	90
5.66	8	80
4.95	7	70
4.24	6	60
3.54	5	50
2.83	4	40
2.12	3	30
1.41	2	20
0.707	1	10

input resistance for the voltmeter in Fig. 6-24. Remember $V_{FS} = 7.07$ V rms and $I_{FS} = 50$ μA.

$S_{hw} = $ _____ $R_{in} = $_____

Step 2. Calculate the value of R_s in Fig. 6-24.

$R_s = R_{in} - R_m = $_____

Step 3. Build the circuit shown in Fig. 6-24. Adjust the output of the function generator to 10 V peak (7.07 V rms) at 1 kHz.

Step 4. Measure the deflection produced for peak values of 10, 9, 8, 7, 6, 5, 4, 3, 2, and 1 V. Complete Table 6-2.

The right-hand column in Table 6-2 indicates the degree of agreement between expected and actual deflections. As you can see, the

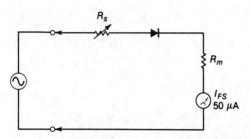

Fig. 6-24. Half-wave, rectifier-type, ac voltmeter for Experiment 6-2.

Table 6-2. Initial Data of Experiment 6-2

V_m (Volts)	D_{meas} (Percent)	$(D_{meas}/D_{scale}) \times 100$
10		
9		
8		
7		
6		
5		
4		
3		
2		
1		

degree of agreement is *not* very good. Notice that the largest error occurs for small input voltages. Recall that when you designed the voltmeter in Fig. 6-24 you employed equations that assumed an ideal diode. One of the conditions that must be met for a real diode to approximate an ideal diode is specified by Equation 6-3, that is, $V >> V_K$. This condition is certainly *not met* for *small* input voltages. To improve the performance of the voltmeter, we will calculate a new value for R_s that takes into account the diode's knee voltage.

The effective peak voltage (V_{mE}) is less than the input peak voltage (V_m) by an amount equal to the diode's knee voltage. Thus,

$$V_{mE} = V_m - V_K$$

$$V_{rms} = 0.707V_{mE} = V_{FS}$$

163

$$R_{\text{in}} = S_{hw}V_{FS}$$
$$= \frac{0.45}{I_{FS}}(0.707V_{mE})$$
$$R_s = R_{\text{in}} - R_m$$

Now,

$$V_m = 10 \text{ V peak}$$
$$V_K = 0.7 \text{ V} \quad (\text{silicon diode})$$
$$I_{FS} = 50 \ \mu\text{A}$$
$$R_m = 5 \text{ k}\Omega \quad (\text{typical value})$$

Substituting values gives

$$V_{mE} = V_m - V_K = 10 \text{ V} - 0.7 \text{ V} = 9.3 \text{ V}$$
$$V_{\text{rms}} = 0.707V_{mE} = 0.707(9.3) = 6.58 \text{ V}$$
$$R_{\text{in}} = \frac{0.45}{I_{FS}}(0.707V_{mE})$$
$$= \frac{0.45}{50 \ \mu\text{A}}(6.58 \text{ V}) = 59.2 \text{ k}\Omega$$
$$R_s = R_{\text{in}} - R_m = 59.2 \text{ k}\Omega - 5 \text{ k}\Omega = 54.2 \text{ k}\Omega$$

Step 5. Using a procedure similar to the one discussed above, calculate a value for R_s that takes the diode's knee voltage into account.
$R_s = $ _____

Step 6. Adjust the value of R_s in your circuit to the value calculated above. Adjust the output of the function generator to 10 V peak. If necessary, further adjust R_s so that full-scale deflection occurs.

Step 7. Repeat Step 4 and complete Table 6-3.

The degree of agreement between expected and actual deflections should be much better than your initial data in Table 6-2. Notice

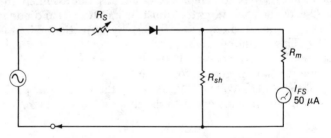

Fig. 6-25. Compensating for small ac voltages in Experiment 6-2.

Table 6-3. Data for New Value of R_s

V_m (Volts)	D_{meas} (Percent)	$(D_{meas}/D_{scale}) \times 100$
10		
9		
8		
7		
6		
5		
4		
3		
2		
1		

again, however, that the largest error occurs for small input voltages. This is because the diode is operating in its nonlinear region. Many commercial instruments provide a separate low-voltage ac scale because of the nonlinear properties of real diodes.

Table 6-4. Data for the Voltmeter in Fig. 6-26

V_m (Volts)	D_{meas} (Percent)	$(D_{meas}/D_{scale}) \times 100$
10		
9		
8		
7		
6		
5		
4		
3		
2		
1		

Step 8. In Fig. 6-25 $R_{sh} = R_m$. Thus the series equivalent of the shunted movement acts as a meter movement whose $I_{FS} = 100\ \mu$A, and whose $R_m' = R_m/2$. Design the voltmeter in Fig. 6-25. The knee voltage of the diode *should be* considered in your design. Build the circuit and complete Table 6-4.

Conclusion

When is it necessary to consider the effect of a diode's knee voltage in a rectifier-type ac voltmeter? If it becomes necessary to measure small ac voltages, what modifications would you suggest for a rectifier-type ac voltmeter?

CHAPTER 7

Peak-Type AC Voltmeters

7-1 INTRODUCTION

Rectifier-type ac voltmeters are calibrated based on the assumption that *only sinusoidal* voltages will be measured. Normally, they are calibrated to indicate the *rms value* of the sinusoidal input voltage. Thus, if you attempt to measure nonsinusoidal input voltages with a rectifier-type ac voltmeter, the reading obtained will probably be very inaccurate. The only *exception* to this general principle occurs for the *special case* where the rms value of the nonsinusoidal waveform *is* $0.707V_m$—the same as a sine wave.* Nonsinusoidal voltages are frequently encountered in communications and other areas of electronics. As you know, *true-rms* instruments *do* indicate the rms value of nonsinusoidal waveforms. If you wanted to measure the harmonic content of an audio signal in a high-fidelity system, you would probably use some type of true-rms instrument for the measurement. Recall, however, that specialized analog instruments, such as true-rms voltmeters, tend to be more expensive than the general-purpose instruments we have been discussing. In less critical applications where nonsinusoidal waveforms are encountered, peak-type ac voltmeters are often used.

7-2 OBJECTIVES

At the end of this chapter you will be able to do the following:

---

* Assuming that a full-wave, rectifier-type, ac voltmeter is used for the measurement.

- Design a half-wave, peak-type, ac voltmeter.
- Design a full-wave, peak-type, ac voltmeter.
- Determine the accuracy of a peak-type voltmeter.
- Design a peak-to-peak type ac voltmeter.
- Discuss the operation of a typical vom.

7-3 BASIC CONCEPTS

Peak-type ac voltmeters utilize one or more diodes to obtain a pulsating dc current similar to the rectifier-type voltmeters discussed in Chapter 6. In a peak-type ac voltmeter, however, the pulsating dc current does *not* flow through the meter movement—it is used instead to charge a capacitor to the *peak value* of the signal being measured. In effect, the ac input voltage is converted to a dc voltage equal to the peak value of the ac voltage. The dc capacitor voltage can then be *measured with a dc voltmeter*. Since peak-type voltmeters are normally calibrated to indicate peak values, the signal being measured *does not* have to be sinusoidal to obtain an accurate reading. Thus a peak-type ac voltmeter would read 5 V peak if the signal being measured was a 5-V-peak sinusoid, square wave, triangular wave, or some other voltage whose peak value was 5 V.

7-4 HALF-WAVE, PEAK-TYPE VOLTMETERS

To begin to understand how half-wave, peak-type, ac voltmeters work, consider the half-wave rectifier in Fig. 7-1. Since the diode only permits current to flow through it in one direction, the output

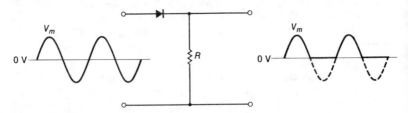

Fig. 7-1. A half-wave rectifier.

voltage will have the same "shape" as the pulsating dc current. Next, consider the circuit illustrated in Fig. 7-2. Note that the resistor in Fig. 7-1 has been replaced with a capacitor in Fig. 7-2. During the first quarter-cycle the diode is forward biased. Thus the capacitor charges to the peak value (V_m) of the input voltage. As soon as the input voltage goes below V_m the voltage across the capacitor reverse biases the diode. Since the input voltage will never exceed V_m the

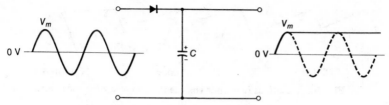

Fig. 7-2. A half-wave peak detector with $R = \infty$.

diode remains reverse biased! Thus, after the first quarter-cycle the voltage across the capacitor is a "pure" dc voltage equal to the peak value of the input voltage. For this reason the circuit in Fig. 7-2 is often called a *peak detector*.

The circuit in Fig. 7-3 is the peak detector of Fig. 7-2 with a resistive load connected across the capacitor. Again, during the first quarter-cycle the capacitor charges to the peak value of the input voltage. Now, however, once the diode becomes reverse biased the capacitor starts to discharge through R. The capacitor voltage continues to decrease, as shown in Fig. 7-3, until the input voltage is once again large enough to forward bias the diode, thus repeating the process just described. If the capacitor voltage does not decrease significantly during the time the diode is reverse biased, then the output voltage in Fig. 7-3 will approximate the pure dc output voltage in Fig. 7-2. As you will soon see, whether or not this is the case will depend on the relationship that exists between the time constant (RC) and the frequency of the input voltage.

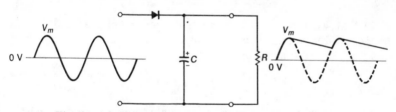

Fig. 7-3. A half-wave peak detector with a resistive load (R).

Fig. 7-4 illustrates the output voltage in Fig. 7-3 for small, medium, and large values of time constant ($\tau = RC$). We will now derive some very useful relationships for the half-wave, peak-type, ac voltmeter. A review of RC charging and discharging circuits is provided in Appendix D for those readers who wish to review this material prior to studying our derivation.

Consider Fig. 7-5, which illustrates in some detail the output voltage in Fig. 7-3. From our knowledge of discharging RC circuits, we can write the following equation to describe the output voltage $v_o(t)$ during the time the capacitor is discharging:

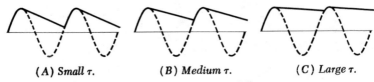

(A) Small τ.   (B) Medium τ.   (C) Large τ.

Fig. 7-4. Variations in output voltage for different time constants.

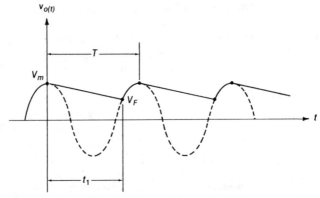

Fig. 7-5. Charging and discharging cycles for the half-wave peak detector in Fig. 7-3.

$$v_0(t) = V_m e^{-t/\tau}$$

At time t_1 voltage $v_0(t)$ has a value of V_F. Thus

$$V_F = V_m e^{-t_1/\tau}$$

Note that V_m represents the *initial* capacitor voltage, and V_F the *final* capacitor voltage. It is useful at this point to define the accuracy of a peak detector as follows:[*]

$$a = \frac{V_F}{V_m} = e^{-t_1/\tau}$$

In a well-designed peak detector (the *only* type we will design) the final voltage (V_F) will be very close in value to the initial voltage (V_m). For this case you can see in Fig. 7-5 that the time interval t_1 is closely approximated by the period (T) of the input voltage. Thus

$$a = \frac{V_F}{V_m} = e^{-T/\tau}$$

A function of the form e^x is closely approximated by $1 + x$ for small values of x. Thus

---

[*] *Cf.* Albert Paul Malvino, *Electronic Instrumentation Fundamentals*, New York: McGraw Hill, 1967, p. 99. This derivation is based on a similar one provided in Dr. Malvino's excellent text.

$$a = 1 - \frac{T}{\tau} \tag{7-1}$$

where

  a = accuracy of a half-wave peak detector,
  T = period of the input voltage,
  τ = time constant = RC.

Since we are accustomed to thinking of accuracy (a) in the form $a = 1 - e$ we will define the error (e) associated with a half-wave peak detector as

$$e = \frac{T}{\tau} \tag{7-2}$$

Solving Equation 7-1 for the time constant τ,

$$\frac{T}{\tau} = 1 - a$$

$$T = (1 - a)\tau$$

$$\tau = \frac{T}{1 - a}$$

Substituting RC for τ and e for $1 - a$ yields

$$RC = \frac{T}{e} \tag{7-3}$$

Equation 7-3 tells us that good accuracy results in a half-wave peak detector when the time constant (RC) is large compared to the period of the input voltage. Remember, good accuracy means the error is small. As you can see in Equation 7-3 a small value of e dictates that the time constant RC must be large compared to the period, T. From Equation 7-3 Table 7-1 is generated which indicates the relationship between τ and T for good ($a \geq 90$ percent) accuracy of a half-wave peak detector.

Fig. 7-6 illustrates a half-wave, peak-type, ac voltmeter. This circuit is quite similar to the half-wave peak detector in Fig. 7-3 which was just analyzed in detail. However, resistance R in Fig. 7-3 has been replaced with a dc voltmeter. As far as the diode and capacitor in Fig. 7-6 are concerned, the input resistance of the dc voltmeter ($R_s + R_m$) acts like the resistance R in Fig. 7-3.

EXAMPLE 7-1

  Design a half-wave, peak-type, ac voltmeter that meets the following specifications:

0- to 10-V range
Utilizes a 100-μA, 1-kΩ movement

Table 7-1. Relative Size of the Time Constant τ (RC) and Period for Good Accuracy (Half-Wave Circuit)

Percent Accuracy ($a \times 100$)	Percent Error ($e \times 100$)	$\dfrac{\tau}{RC} = \dfrac{T}{e}$
99	1	$100T$
98	2	$50T$
97	3	$33.3T$
96	4	$25T$
95	5	$20T$
94	6	$16.7T$
93	7	$14.3T$
92	8	$12.5T$
91	9	$11.1T$
90	10	$10T$

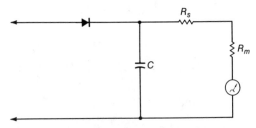

Fig. 7-6. Half-wave, peak-type, ac voltmeter.

Has an accuracy $a = 95$ percent

Can be used to measure signals in the audio-frequency range (20 Hz to 20 kHz)

To begin, the values of R_{in}, and R_s for the dc voltmeter portion of the instrument must be calculated. The sensitivity is

$$S = \frac{1}{I_{FS}} = \frac{1}{100 \ \mu\text{A}} = 10 \ \frac{\text{k}\Omega}{\text{V}}$$

and

$$R_{\text{in}} = SV_{FS}$$
$$= \left(10 \ \frac{\text{k}\Omega}{\text{V}} \right) (10 \ \text{V}) = 100 \ \text{k}\Omega$$

The series resistance is

$$R_s = R_{\text{in}} - R_m$$
$$= 100 \ \text{k}\Omega - 1 \ \text{k}\Omega$$
$$= 99 \ \text{k}\Omega$$

From Equation 7-3 or Table 7-1 you know that for an accuracy of 95 percent

$$RC = 20T$$

Clearly the value of R to use in this equation is 100 kΩ—the input resistance (R_{in}) of the dc voltmeter. The specifications indicate that the voltmeter should be able to measure voltages whose frequency can be as low as 20 Hz or as high as 20 kHz. When a range of frequencies is stated in the specifications for the voltmeter, you should design the voltmeter to provide the specified accuracy (95 percent in this example) at the *minimum* frequency. This ensures that accuracies for frequencies higher than the minimum frequency will *exceed* the accuracy obtained *at* the minimum frequency. Since period and frequency are reciprocals, the appropriate value of T to use is T_{max}. Thus

$$T_{max} = \frac{1}{f_{min}} = \frac{1}{20 \text{ Hz}} = 0.05 \text{ s}$$

and

$$R_{in}C = 20T_{max}$$

or

$$(100 \text{ k}\Omega)C = 20(0.05 \text{ s})$$

$$C = \frac{20(0.05) \text{ s}}{100 \text{ k}\Omega} = 10 \text{ } \mu\text{F}$$

The design is illustrated in Fig. 7-7.

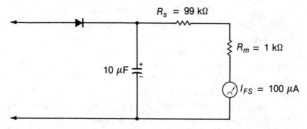

Fig. 7-7. Half-wave, peak-type, ac voltmeter for Example 7-1.

EXAMPLE 7-2

Indicate how the meter face for the voltmeter in Example 7-1 should be calibrated assuming the voltmeter is used to measure:
  (a) Peak values for voltages of *any* shape.
  (b) The rms values of sawtooth voltages.
  (a) The voltmeter was designed so that full-scale deflection occurs when the input voltage has a *peak value* of 10 V. The shape of the input voltage is *not* important in a peak-type ac voltmeter, *assuming* that the meter face is calibrated to read *peak* values. Thus an input voltage whose peak value is 10 V will produce full-scale deflection. Similarly, an input volt-

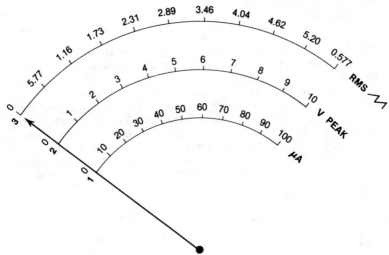

Fig. 7-8. Calibrated meter face for the voltmeter in Fig. 7-7.

age whose peak value is 5 V will produce half-scale deflection, and so on. The calibrated meter face is shown in Fig. 7-8. The μA (first) scale simply shows how the 0- to 100-μA meter scale is divided in ten equal parts. The second scale in Fig. 7-8 indicates how the meter face would be calibrated to read peak values.

(b) Normally peak-type voltmeters are calibrated to indicate the peak value of the voltage being measured. As you know, rectifier-type voltmeters are normally calibrated to indicate the rms value of *sinusoidal* input voltages. This example problem is cleverly designed to illustrate the fact that *you can calibrate your instruments to meet your specific needs!* In this example it has been assumed that a need exists to measure the rms value of sawtooth voltages. Of course, a true-rms instrument could be used for such measurements, but if your application does not demand precise measurements, the voltmeter in Example 7-1 could be used once the meter has been properly calibrated.

Fig. 7-9 illustrates a sawtooth voltage waveform. As you can see the rms value of this sawtooth voltage is approximately $0.577V_m$. Thus, by multiplying the peak values (scale 2) in Fig. 7-8 by 0.577, the corresponding rms values for the sawtooth voltage are obtained. This is illustrated by the rms (third) scale in Fig. 7-8.

EXAMPLE 7-3

For the voltmeter in Fig. 7-10 determine:
   (a) The peak input voltage required for full-scale deflection.
   (b) The lowest frequency that an input voltage can have if the minimum acceptable accuracy is 90 percent.

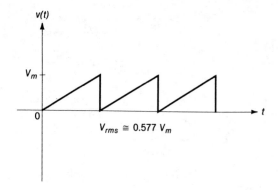

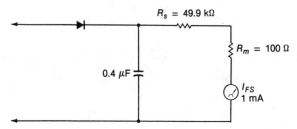

Fig. 7-9. Sawtooth voltage for Example 7-2.

Fig. 7-10. Half-wave, peak-type voltmeter for Example 7-3.

(c) Assuming the voltmeter is used to measure the voltage between A and B in Fig. 7-11, estimate the reading of the voltmeter.

(a) For full-scale deflection $I_m = I_{FS} = 1$ mA. The voltage across the capacitor equals the voltage across the series combination of R_s and R_m. Thus

$$V_C = (1 \text{ mA})(49.9 \text{ k}\Omega + 100 \text{ }\Omega) = 50 \text{ V}$$

Since the voltage across C equals V_m, the peak input voltage is $V_m = 50$ V.

(b) Recall that "worst case" accuracy occurs at the low-frequency limit for which the instrument was designed. From Equation 7-3 or Table 7-1, for an accuracy of 90 percent,

Fig. 7-11. Circuit for Example 7-3.

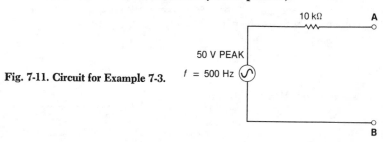

$$R_{in}C = 10T_{max}$$

Thus

$$T_{max} = \frac{R_{in}C}{10} = \frac{(50\ k\Omega)(0.4\ \mu F)}{10}$$
$$= 2\ ms$$

$$f_{min} = \frac{1}{T_{max}} = \frac{1}{2\ ms} = 0.5 \times 10^3\ Hz$$
$$= 500\ Hz$$

(c) Recall that the definition for the accuracy of a peak-type voltmeter is

$$a = \frac{V_F}{V_m} = 1 - \frac{T}{\tau}$$

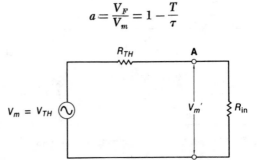

Fig. 7-12. A generalized Thevenin equivalent circuit for Example 7-3.

It is important to realize that this definition *does not* include any loading effects. The purpose of this portion of the problem is to demonstrate how loading effects are calculated for peak-type voltmeters.

In Fig. 7-11 the peak voltage between A and B is clearly 50 V. For all practical purposes the voltmeter has an input resistance of 50 kΩ. When the voltmeter is placed across AB the 50-V peak input voltage divides between the 10-kΩ resistance and the 50-kΩ input resistance of the voltmeter. Thus, because of loading, the peak voltage between A and B decreases when the voltmeter is placed in the circuit. Once again, in order to obtain very general results we will employ Thevenin's theorem. Referring to Fig. 7-12,

$$V_m' = V_m\left(\frac{R_{in}}{R_{TH} + R_{in}}\right)$$

where V_m' is the peak voltage between A and B *when* the voltmeter *is* connected. Defining the ratio

$$\frac{R_{in}}{R_{TH} + R_{in}}$$

as k an accuracy equation can be written which *includes loading effects* for a peak-type ac voltmeter:

$$a' = \left(\frac{R_{in}}{R_{TH} + R_{in}}\right)\left(1 - \frac{T}{\tau}\right) \qquad (7\text{-}4)$$

or

$$a' = ka$$

Returning to the problem at hand,

$$k = \frac{R_{in}}{R_{TH} + R_{in}} = \frac{50\text{ k}\Omega}{10\text{ k}\Omega + 50\text{ k}\Omega} = \frac{50\text{ k}\Omega}{60\text{ k}\Omega} = 0.833$$

$$T = \frac{1}{500\text{ Hz}} = 2\text{ ms}$$

$$\tau = R_{in}C = (50\text{ k}\Omega)(0.4\ \mu\text{F}) = 20\text{ ms}$$

Thus

$$1 - \frac{T}{\tau} = 1 - \frac{2\text{ ms}}{20\text{ ms}} = 1 - 0.1 = 0.9$$

Hence

$$a' = 0.833(0.9) = 0.75$$

$$V_m' = kV_m = 0.833(50\text{ V}) = 41.7\text{ V}$$

The final capacitor voltage is

$$V_F = a'V_m = 0.75(50\text{ V}) = 37.5\text{ V}$$

The voltage across the capacitor is illustrated in Fig. 7-13. The dc voltmeter will deflect to the average value of the voltage in Fig. 7-13. This average value can be estimated with the equation

$$V_{avg} = \frac{V_m' + V_F}{2}$$

Thus

$$V_{avg} = \frac{41.7\text{ V} + 37.5\text{ V}}{2} = 39.6\text{ V}$$

7-5 FULL-WAVE, PEAK-TYPE VOLTMETERS

A full-wave, peak-type, ac voltmeter is illustrated in Fig. 7-14. Like the half-wave version the basic idea is to charge C to the peak value of the input voltage, and then measure the voltage across C with a dc voltmeter. Note in Fig. 7-15 that the unfiltered (dotted lines) output voltage from a full-wave rectifier has a frequency which is *twice* the input frequency. Thus the period of the output voltage is one-half the period of the input voltage. In a manner simi-

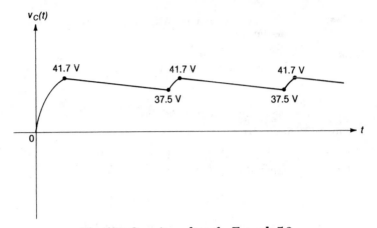

Fig. 7-13. Capacitor voltage for Example 7-3.

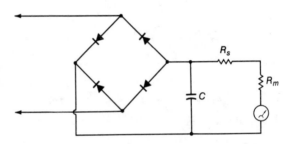

Fig. 7-14. A typical full-wave, peak-type, ac voltmeter.

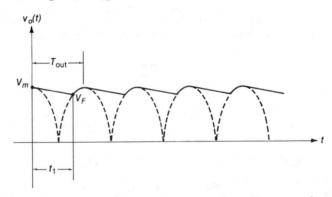

Fig. 7-15. Charging and discharging cycles for the full-wave, peak-type, ac voltmeter in Fig. 7-14.

lar to the analysis provided for the half-wave, peak-type voltmeter we will now derive the analogous relationships for the full-wave version. With reference to Fig. 7-15,

$$v_0(t) = V_m\,e^{-t/\tau}$$

$$V_F = V_m\,e^{-t_1/\tau}$$

$$a = \frac{V_F}{V_m} = e^{-t_1/\tau}$$

Since $t_1 \cong T/2$ we have

$$a \cong e^{-T/2\tau}$$

where T is the period of the *input* voltage. Employing the now familiar small-argument approximation, $e^x = 1 + x$, we have

$$a = 1 - \frac{T}{2\tau} \tag{7-5}$$

The error e is thus defined as

$$e = \frac{T}{2\tau} \tag{7-6}$$

Solving Equation 7-6 for τ yields

$$2\tau e = T$$

$$\tau = \frac{T}{2e} \tag{7-7}$$

Equation 7-7 is used to generate Table 7-2, which shows the relationship between τ and T for good accuracy ($a \geqq 90$ percent) of a full-wave peak detector.

Table 7-2. Relative Size of Time Constant τ (RC) and Period (T) for Good Accuracy (Full-Wave Circuit)

Percent Accuracy ($a \times 100$)	Percent Error ($e \times 100$)	$RC = \dfrac{\tau}{T} = \dfrac{T}{2e}$
99	1	50T
98	2	25T
97	3	16.7T
96	4	12.5T
95	5	10T
94	6	8.3T
93	7	7.1T
92	8	6.25T
91	9	5.6T
90	10	5T

To include loading effects for a full-wave, peak-type voltmeter we can use

$$a' = \left(\frac{R_{in}}{R_{TH} + R_{in}}\right)\left(1 - \frac{T}{2\tau}\right) \qquad (7\text{-}8)$$

or $a' = ka$.

EXAMPLE 7-4

Design a full-wave, peak-type voltmeter to measure voltages up to 50 V peak. Assume that a 1-mA, 100-Ω movement is available for the design and that the lowest frequency to be measured is 500 Hz. The accuracy of the voltmeter should equal or exceed 90 percent.

Notice that the specifications for the voltmeter are identical with those for the voltmeter discussed in Example 7-3. Thus the dc portion of the voltmeter will be the same as in Example 7-3. Consequently,

$$R_s = 49.9 \text{ k}\Omega$$

From Table 7-2 you can see that for an accuracy of 90 percent,

$$RC = 5T$$

Thus

$$R_{in}C = 5T_{max}$$

so that

$$C = \frac{5T_{max}}{R_{in}} = \frac{5(1/500 \text{ Hz})}{50 \text{ k}\Omega}$$

$$C = \frac{5(2 \text{ ms})}{50 \text{ k}\Omega} = 0.2 \ \mu\text{F}$$

Comparing Examples 7-3 and 7-4 indicates that for identical specifications a full-wave, peak-type voltmeter requires *half the capacitance* of a half-wave, peak-type voltmeter. The final design is illustrated in Fig. 7-16.

7-6   PEAK-TO-PEAK VOLTMETERS

The peak detectors discussed previously were examples of positive peak detectors. As you know, a positive peak detector provides a dc output voltage equal to the positive peak of the input voltage. By *reversing* the diode(s) of a positive peak detector you obtain a negative peak detector. A negative peak detector provides a dc output voltage equal to the negative peak of the input voltage.

Fig. 7-17A illustrates a positive peak detector. Recall that during the first quarter cycle the diode is forward biased. Thus, C charges to V_m with the polarity shown in Fig. 7-17A. As a result $v_o = + V_m$. By reversing the diode in Fig. 7-17A you obtain the negative peak detector illustrated in Fig. 7-17B. Note in Fig. 7-17B that the diode

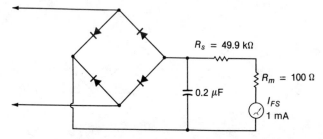

Fig. 7-16. Full-wave, peak-type voltmeter for Example 7-4.

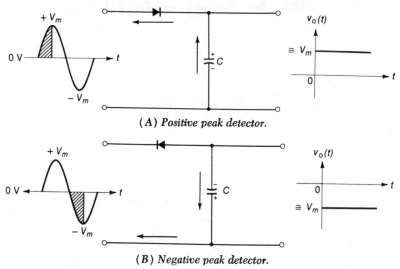

(A) Positive peak detector.

(B) Negative peak detector.

Fig. 7-17. Peak detectors.

now becomes forward biased during the third quarter cycle. Thus, the direction of current in Fig. 7-17B is just the opposite of Fig. 7-17A. As a result, the capacitor in Fig. 7-17B charges so that $v_o = -V_m$.

EXAMPLE 7-5

Estimate the capacitor voltage in Fig. 7-17A (positive peak detector) and in Fig. 7-17B (negative peak detector) if the input voltage is the waveform shown in: (a) Fig. 7-18A, (b) Fig. 7-18B, and (c) Fig. 7-18C.

Ideally, the output (capacitor) voltage of a positive peak detector is a dc voltage equal to the positive peak of the input voltage. Similarly, the output voltage of a negative peak detector is a dc voltage equal to the negative peak of the input voltage. By the process of "astute observation" (looking) we can construct Table 7-3.

Thus the capacitor voltage in Fig. 7-17A (positive peak detector) would be 10 V for *any* of the waveforms shown in Figs.

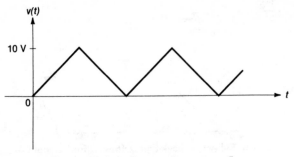

(A) Triangular voltage with positive offset.

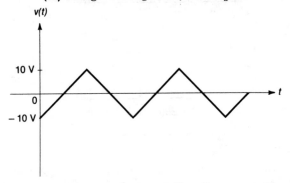

(B) Symmetrical triangular voltage.

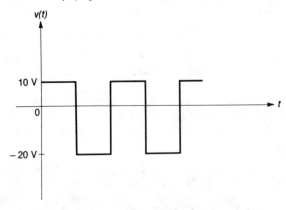

(C) Square wave voltage with negative offset.

Fig. 7-18. Input voltage waveforms.

7-18A, 7-18B, and 7-18C. The capacitor voltage in Fig. 7-17B (negative peak detector) would be 0 V for the waveform shown in Fig. 7-18A, −10 V for the waveform shown in Fig. 7-18B, and −20 V for the waveform shown in Fig. 7-18C.

Table 7-3. Astute Observations for Positive and Negative Peak Voltages (Example 7-5)

Figure	Positive Peak	Negative Peak
7-18A	10 V	0 V
7-18B	10 V	−10 V
7-18C	10 V	−20 V

Clamping circuits change the average value of a waveform. An example of a positive clamp is illustrated in Fig. 7-19A. Note that the positive clamp in Fig. 7-19A is essentially a peak detector with the output taken across the *diode* rather than the capacitor. Recall that once the capacitor has charged to V_m the diode acts like an open circuit. Ideally, the charged capacitor acts like a battery whose terminal voltage is equal to V_m. Thus we can picture the equivalent

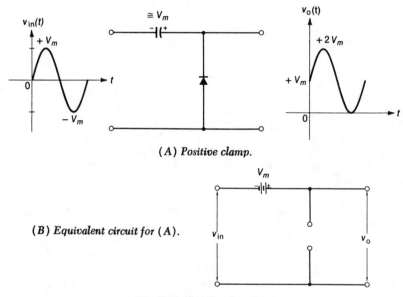

(A) Positive clamp.

(B) Equivalent circuit for (A).

Fig. 7-19. Positive clamping.

circuit illustrated in Fig. 7-19B for the positive clamp in Fig. 7-19A. Going around the loop in Fig. 7-19B we see that the output voltage (v_o) is

$$v_o = V_m + v_{in} \qquad (7\text{-}9)$$

Equation 7-9 indicates that a dc voltage equal to V_m is added to v_{in} to produce v_o. The effect of this addition is illustrated by the output voltage in Fig. 7-19A. Note that the output voltage in Fig. 7-19A has

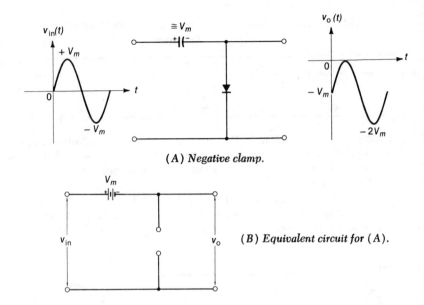

(A) Negative clamp.

(B) Equivalent circuit for (A).

Fig. 7-20. Negative clamping.

an average value equal to V_m. Also, note that the input voltage in Fig. 7-19A has an average value of zero.

By reversing the diode in Fig. 7-19A, you obtain the negative clamp illustrated in Fig. 7-20A. The capacitor in Fig. 7-20A (negative clamp) charges to V_m with a polarity that is just the opposite of the capacitor in Fig. 7-19A (positive clamp). The equivalent circuit for the negative clamp in Fig. 7-20A is illustrated in Fig. 7-20B. Going around the loop in Fig. 7-20B we see that the output voltage (v_o) is

$$v_o = v_{in} - V_m \qquad (7\text{-}10)$$

Equation 7-10 indicates that a dc voltage equal to V_m is *subtracted* from v_{in} to produce v_o. The effect of this subtraction is illustrated by the output voltage in Fig. 7-20A. Note that the output voltage in Fig. 7-20A has an average value of $-V_m$, while the input voltage has an average value of zero.

A peak-to-peak detector, also known as a *voltage doubler*, provides a dc output voltage equal to the peak-to-peak value of the input voltage. An example of a positive peak-to-peak detector is illustrated in Fig. 7-21. As you can see, the circuit consists of a positive clamp (C_1 and D_1) followed by a positive peak detector (C_2 and D_2). Note that the output of the positive clamp is the input to the positive peak detector. Since the output of the positive clamp has a

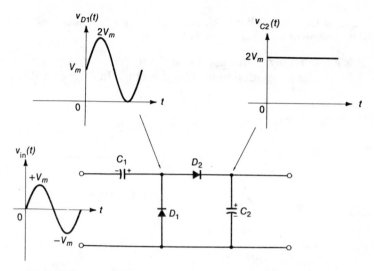

Fig. 7-21. A positive peak-to-peak detector (voltage doubler).

positive peak equal to $2V_m$ the output of the positive peak detector will be a dc voltage equal to $+2V_m$. If *both* diodes in Fig. 7-21 were reversed we would have a negative clamp followed by a negative peak detector. In such a case the dc output voltage would equal $-2V_m$. The dc output voltage would be zero if *just* D_1 or D_2 were reversed. When building peak-to-peak detectors you must make sure the diodes are placed in the circuit correctly!

Fig. 7-22 illustrates a peak-to-peak type ac voltmeter. As you can see we have simply added a dc voltmeter to measure the dc voltage across C_2 in the peak-to-peak detector of Fig. 7-21. Naturally, a peak-to-peak type ac voltmeter is *normally* calibrated to indicate the peak-to-peak value of the input voltage.

EXAMPLE 7-6

Sketch the waveforms appearing at the output of the positive clamp (v_{D1}) and the positive peak detector (v_{C2}) in Fig. 7-22 if the input voltage is the waveform shown in Figs. 7-18A, 7-18B, and 7-18C.

Fig. 7-23 illustrates the waveforms for this example. The input voltages (Figs. 7-18A, 7-18B, and 7-18C) are redrawn in Fig. 7-23 in the INPUT VOLTAGE column. For each input voltage the capacitor associated with the positive clamp (C_1 in Fig. 7-22) charges to a value equal to the *peak* of the input voltage. This dc voltage is then "added" to the input voltage to produce the clamped output waveforms $(v_{D1}(t))$ shown in Fig. 7-23. The clamped output voltages are then peak detected by D_2 and C_2 in Fig. 7-22. Thus, the voltage across C_2 in Fig. 7-22 will be a dc voltage equal to the *peak-to-peak*

185

value of the input voltage. This dc voltage is easily measured with a dc voltmeter as indicated in Fig. 7-22.

It is worth noting at this point that the design of clamping circuits is quite similar to the design of peak detectors. In each case it is de-

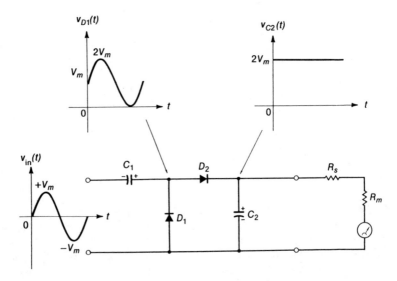

Fig. 7-22. A peak-to-peak type ac voltmeter.

sirable to make the time constant (RC) large compared to the period of the input voltage. This practice will help to ensure good clamping and peak detection. Thus, in many peak-to-peak type ac voltmeters you will find $C_1 = C_2$. The value of C_2, of course, is determined by the lowest input frequency and desired accuracy.

The *exact* relationships between frequency, accuracy, loading, and the time required for C_1 and C_2 in Fig. 7-22 to charge to their final dc voltages is quite complex. Most of the time the simplified relationships developed previously are sufficient to design functional clamps and peak detectors. Occasionally, however, due to transients and a generally pervasive universe, problems may arise. For example, if the input voltage in Fig. 7-22 has a negative peak of 0 V, then the final voltage across C_1 should be 0 V. This concept (and others) was illustrated in Example 7-6. In practice, however, it may take an excessive amount of time for the voltage across C_1 to read the final value of 0 V. Obviously, this condition is undesirable. One possible solution to this problem is to shunt D_1 with a resistance. The value of this resistance can best be determined experimentally.

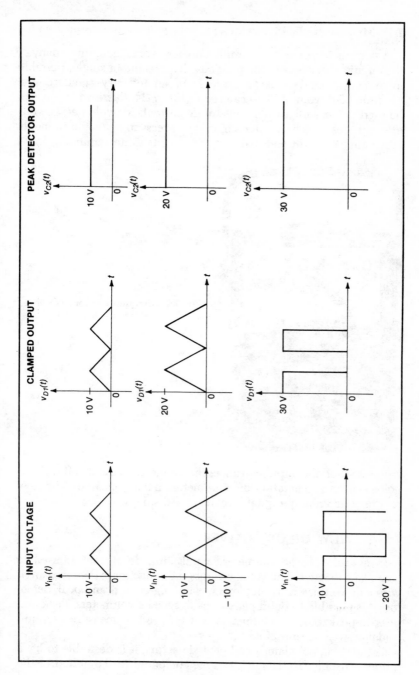

Fig. 7-23. Waveforms for Example 7-6.

7-7 MULTIMETERS (VOM'S)

A multimeter, or volt-ohm-milliammeter (vom), combines many of the circuits discussed in the previous chapters into a single, portable instrument. Fig. 7-24 is an example of an industry standard—the Simpson 260® vom. As you can see in Fig. 7-24, this instrument can be used to measure a wide range of dc currents, dc voltages, ac voltages, and resistance. If you own or have access to a vom, examine its schematic. You will find many of the circuits quite familiar!

Fig. 7-24. Simpson 260 vom. (*Courtesy Simpson Electric Co.*)

Because of the rapid advances in integrated-circuit (IC) technology, a new generation of multimeters have appeared. The new multimeters employ digital circuits and displays.

7-8 REVIEW OF OBJECTIVES

In a peak-type ac voltmeter, a pulsating dc current is used to charge a capacitor. The voltage across the capacitor is then measured with a dc voltmeter. By employing clamping circuits and peak detectors it is possible to build peak-to-peak type ac voltmeters. Peak and peak-to-peak type ac voltmeters can be used to measure nonsinusoidal as well as sinusoidal voltages.

In the design of clamps and peak detectors, it is desirable to have a time constant (RC) which is large compared to the period of the input voltage.

Multimeters, or vom's, are instruments which combine the circuits previously discussed into a single portable instrument.

7-9 QUESTIONS

1. What is the relationship between input frequency and time constant for good peak detection?
2. If a peak detector has an accuracy of 95 percent at 100 Hz, will the accuracy be higher or lower at 1 kHz? Why?
3. What is the difference between a positive clamp and a negative clamp?
4. What is a peak-to-peak detector?
5. Where would a peak-type voltmeter be preferred to a rectifier-type voltmeter?

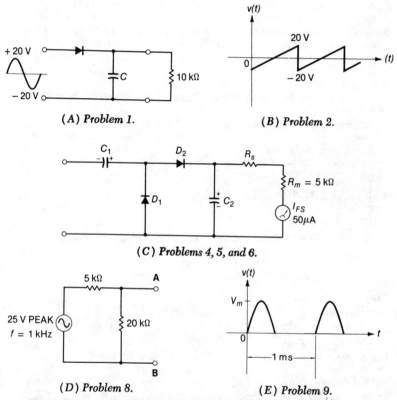

(A) Problem 1.

(B) Problem 2.

(C) Problems 4, 5, and 6.

(D) Problem 8.

(E) Problem 9.

Fig. 7-25. Waveforms and schematic diagrams for Chapter 7 problems.

6. What is another name for a peak-to-peak detector?
7. In an attempt to build a "voltage doubler," Dr. Courtine cascaded the output of a positive clamp and negative peak detector. What output voltage did the good doctor measure? Why?

7-10 PROBLEMS

1. The peak detector in Fig. 7-25A is designed to measure voltages between 100 Hz and 50 kHz. What is the minimum value for C, assuming the accuracy should be at least 90 percent?

2. The waveform in Fig. 7-25B is applied to the input of a: (a) positive clamp, (b) negative clamp, (c) positive peak detector, (d) negative peak detector, (e) positive peak-to-peak detector, and (f) negative peak-to-peak detector. In each case sketch the output waveform.

3. A full-wave peak detector is designed to measure voltages between 20 Hz and 30 kHz. What size capacitor would you recommend if the desired accuracy is 95 percent? Assume $R = 10$ kΩ.

4. The voltmeter in Fig. 7-25C is calibrated to read peak-to-peak values. If the dc voltmeter reads 10 V, what is the peak input voltage?

5. A 10-V peak input voltage in Fig. 7-25C produces a current of 50 μA in the meter movement. Determine the value of R_s and the voltage across C_2.

6. What values for C_1 and C_2 would you recommend in Problem 5 if the minimum input frequency is 200 Hz? Assume an accuracy of 90 percent.

7. Design a half-wave, peak-type, ac voltmeter for use at 1 kHz. A 100-μA, 2-kΩ movement is available for the design. Assume the desired accuracy is 90 percent and $V_{FS} = 25$ V peak.

8. What will the voltmeter in Problem 7 read if it is connected between A and B in Fig. 7-25D?

9. The rms value of the waveform in Fig. 7-25E is $0.5V_m$. Assume $V_m = 10$ V. Design a peak-type ac voltmeter utilizing a 100-μA, 2-kΩ movement that provides the following scales: peak value and rms value. Indicate how each scale should be calibrated. Assume the desired accuracy is 90 percent.

7-11 EXPERIMENT 7-1

Objective

The objective of this experiment is to investigate clamps and peak detectors.

Material Required

Function generator
Oscilloscope
1N4001 diode (almost any general-purpose diode will do)
10-μF capacitor
Resistor decades, variable in 100-kΩ, 10-kΩ, and 1-kΩ steps

Introduction

Clamps change the *average value* of a waveform. Peak detectors provide a dc output voltage equal to the peak value of the input voltage. Both clamps and peak detectors consist of the series combination of a diode and capacitor. The output voltage of a clamp is taken across the diode, while the output voltage of a peak detector is taken across the capacitor.

In this experiment you will investigate a positive clamp and a positive peak detector. The output voltage of a positive clamp is specified by Equation 7-9:

$$v_o = V_m + v_{in} \qquad (7\text{-}9)$$

Ideally, the output voltage of a positive peak detector is V_m. Recall that the accuracy of a half-wave peak detector is provided by Equation 7-1:

$$a = 1 - \frac{T}{\tau} \qquad (7\text{-}1)$$

For clamps and peak detectors to approximate ideal circuits, one of the conditions which must be satisfied is

$$\tau \gg T$$

If this condition is not satisfied, then the circuits will *not* function properly.

Procedure

Step 1. Build the circuit shown in Fig. 7-26A.

Step 2. Set R to 1 MΩ. Adjust v_{in} to 10 V peak (sinusoidal) at 100 Hz. Measure v_{in} and v_o with an oscilloscope. Make neat sketches of v_{in} and v_o indicating peak values. Repeat for square wave and triangular wave input voltages if available. Record your results in Table 7-4.

Step 3. Repeat Step 2 for the following values of R: 800 kΩ, 600 kΩ, 400 kΩ, 200 kΩ, 100 kΩ, and 10 kΩ.

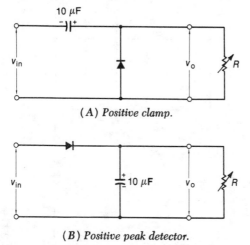

(A) Positive clamp.

(B) Positive peak detector.

Fig. 7-26. Schematic diagrams for Experiment 7-1.

Table 7-4. Data for Circuit of Fig. 7-26A

Quantity	Resistance R							
	1 MΩ	800 kΩ	600 kΩ	400 kΩ	200 kΩ	100 kΩ	10 kΩ	
$v_{in} = 10$ V peak sinusoidal wave v_{in}								
v_o								
square wave v_{in}								
v_o								
triangular wave v_{in}								
v_o								
$v_{in} = 2$ V peak v_{in}								
v_o								
$v_{in} = 1$ V peak v_{in}								
v_o								

Table 7-5. Data for Circuit of Fig. 7-26B

Quantity	Resistance R								
	1 MΩ	800 kΩ	600 kΩ	400 kΩ	200 kΩ	100 kΩ	10 kΩ		
$v_{1n} = 10$ V peak sinusoidal wave v_{1n}									
v_o									
square wave v_{1n}									
v_o									
triangular wave v_{1n}									
v_o									
$v_{1n} = 2$ V peak v_{1n}									
v_o									
$v_{1n} = 1$ V peak v_{1n}									
v_o									

Step 4. Readjust R to 1 MΩ. Reduce v_{in} to 2 V peak (sinusoidal). Measure v_{in} and v_o with an oscilloscope, and sketch the resulting waveforms. Repeat for $v_{in} = 1$ V peak (sinusoidal).

Step 5. Repeat Steps 2 to 4 for the circuit shown in Fig. 7-26B, putting your data in Table 7-5.

Discussion

When $R = 1$ MΩ and $v_{in} = 10$ V peak, both circuits should have provided outputs that were *nearly* ideal. As R was decreased from 1 MΩ to 10 kΩ you probably observed increasingly degraded output waveforms. One reason for this is that the circuit time constant (RC) becomes smaller as R decreases. Recall that RC must be much greater than the period of the input signal for near ideal operation.

When $R = 1$ MΩ and the peak value of the input voltage was "small" (1 or 2 volts) you again observed a degraded output. Remember that you are using *real* diodes which require a driving voltage greater than the diodes' knee voltage for near ideal operation. Thus, if the input voltages are not large compared to the diodes' knee voltage, the circuits will not work properly.

Conclusion

Specify the conditions which must be met for near ideal operation of clamps and peak detectors. If your background includes a knowledge of op-amp circuits, explain how the block diagrams in Figs. 7-27A and 7-27B "solve" some of the nonideal effects encountered in this experiment.

7-12 EXPERIMENT 7-2

Objective

The objective of this experiment is to design a peak-to-peak type ac voltmeter.

Material Required

50-μA meter movement
Function generator
Oscilloscope
Two 1N4001 diodes (almost any general-purpose diodes will do)
Resistor decades, variable in 100-kΩ, 10-kΩ, and 1-kΩ steps
Two capacitors—the values will be calculated as part of the experiment

Introduction

In the previous experiment you investigated the positive clamp and positive peak detector. In this experiment we will employ these

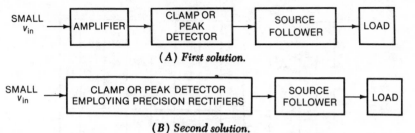

(A) First solution.

(B) Second solution.

Fig. 7-27. Proposed solutions for nonideal diode effects.

circuits and a dc voltmeter to form a peak-to-peak type ac voltmeter. The output of the positive clamp has a positive peak equal to $2V_m$. This signal is converted to dc by the positive peak detector, which is then measured with a dc voltmeter.

Procedure

Step 1. The voltmeter in Fig. 7-28 should provide full-scale deflection when $v_{in} = 10$ V peak. Assuming an input frequency of 100 Hz, calculate the required value of R_s.
$R_s = $ _____ kΩ

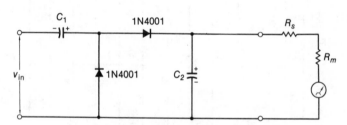

Fig. 7-28. Schematic diagram for Experiment 7-2.

Step 2. Calculate the value of C_2 for an accuracy of 99 percent. To simplify the design process we will make $C_2 = C_1 = C$. Record the calculated value for C.
$C = $ _____ μF.
For this experiment pick a standard value capacitor whose capacitance *equals or exceeds* your calculated value. Limit your choice to values less than 47 μF.

Step 3. Build the circuit shown in Fig. 7-28. Adjust v_{in} to 10 V peak at 100 Hz. If necessary, adjust the value of R_s to obtain full-scale deflection.

Step 4. Monitor the voltage across C_2 with an oscilloscope. Vary the amplitude of v_{in} from 10 V peak to 1 V peak, in 1-V steps. In each case record the reading of the oscilloscope and dc

Table 7-6. Data for Peak-to-Peak Voltmeter

v_{in} peak	Oscilloscope			Voltmeter		
	sinusoidal	square	triangular	sinusoidal	square	triangular
10 V						
9 V						
8 V						
7 V						
6 V						
5 V						
4 V						
3 V						
2 V						
1 V						

voltmeter in Table 7-6. Repeat for triangular wave and square wave inputs if available.

Conclusion

If it was necessary to adjust R_s in Step 3, explain why this was the case. The oscilloscope and dc voltmeter readings tend to disagree when small input voltages are measured. Why? Where could an instrument like the one designed in this experiment be used? Why would it be preferred to a rectifier-type instrument?

CHAPTER 8

DC Bridges, Transducers, and Electronic Instruments

8-1 INTRODUCTION

We have devoted seven lucid chapters in order to develop an understanding of how dc current, dc voltage, resistance, and ac voltage can be measured.

In the real world it is often necessary to measure nonelectrical quantities, such as strain, position, light intensity, and so on. The goal of this chapter is to show you how this can be done, utilizing a few new concepts, and the instruments you are already familiar with!

8-2 OBJECTIVES

At the end of this chapter you will be able to do the following:

- Analyze a Wheatstone bridge circuit.
- Understand how a Wheatstone bridge ohmmeter works.
- Analyze a dual-supply bridge circuit.
- Explain how transducers are used with bridge circuits to measure nonelectrical quantities.
- Design elementary electronic voltmeters, ammeters, and ohmmeters using op-amp circuits.

8-3 THE WHEATSTONE BRIDGE

Fig. 8-1 illustrates a *Wheatstone* bridge. The "null detector" is generally a sensitive ammeter. If the voltage between A and B is zero, then the bridge is said to be *balanced*. If the voltage between

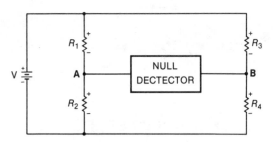

Fig. 8-1. Wheatstone bridge.

A and B is not zero, the bridge is considered *unbalanced*. This condition would be indicated by some deflection of the null detector.

The direction of current through the null detector in Fig. 8-1 is determined by the voltages across R_2 and R_4. If the voltage across R_2 is greater than the voltage across R_4, then point A is more positive with respect to ground than point B. As a result, the electron flow through the null detector is from B to A. Similarly, if the voltage across R_4 is greater than the voltage across R_2, then the current through the null detector is from A to B. Since it is possible for current to flow in either direction through the null detector, most null detectors have their zero position at midscale. When the voltage across R_2 exactly equals the voltage across R_4 voltage $V_{AB} = 0$. In this (balanced) case there would be no deflection of the null detector since the current through it is zero.

In order to see what conditions are necessary to balance a Wheatstone bridge we can employ the equivalent circuit shown in Fig. 8-2. The branch AB in Fig. 8-2 may be considered an open circuit,

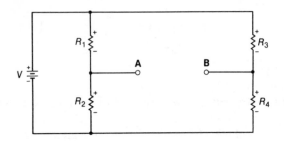

Fig. 8-2. Equivalent circuit for a balanced Wheatstone bridge.

since, when the bridge is balanced, the current through the branch is zero. Thus

$$V_{R2} = V_A = V\left(\frac{R_2}{R_1 + R_2}\right)$$

$$V_{R4} = V_B = V\left(\frac{R_4}{R_3 + R_4}\right)$$

For balance $V_A = V_B$. Therefore

$$V\left(\frac{R_2}{R_1 + R_2}\right) = V\left(\frac{R_4}{R_3 + R_4}\right)$$

and so

$$R_2(R_3 + R_4) = R_4(R_1 + R_2)$$

Multiplying,

$$R_2R_3 + R_2R_4 = R_1R_4 + R_2R_4$$

Finally,

$$R_1R_4 = R_2R_3 \qquad (8\text{-}1)$$

Some readers may prefer the equivalent relationship:

$$\frac{R_1}{R_2} = \frac{R_3}{R_4}$$

Thus the Wheatstone bridge will be balanced when the ratio of the resistances on one side of the bridge equals the corresponding ratio on the other side of the bridge. Alternately, we can state that balance is achieved when the *cross products* of the resistances are equal.

EXAMPLE 8-1

Determine if the bridge in Fig. 8-3 is balanced. If the bridge is not balanced, what value of R_4 is required to balance the bridge?

For balance, $R_1R_4 = R_2R_3$. Substituting the values given in Fig. 8-3 yields

$$10\,\text{k}\Omega(5\,\text{k}\Omega) = 20\,\text{k}\Omega(3\,\text{k}\Omega)$$
$$50\,\text{M}\Omega \neq 60\,\text{M}\Omega$$

Thus with $R_4 = 5\,\text{k}\Omega$ the bridge is *not* balanced.

To determine the value of R_4 required to balance the bridge we solve Equation 8-1 for R_4 and simply substitute the values specified for R_1, R_2, and R_3. Thus

$$R_4 = \frac{R_2R_3}{R_1} = \frac{(20\,\text{k}\Omega)\,(3\,\text{k}\Omega)}{10\,\text{k}\Omega}$$
$$= 6\,\text{k}\Omega$$

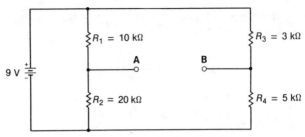

Fig. 8-3. Circuit for Example 8-1.

EXAMPLE 8-2

The bridge in Fig. 8-1 has the following resistance values: $R_1 = R_2 = 10$ kΩ, $R_3 = 20$ kΩ, and $R_4 = 5$ kΩ. Determine the current in the branch AB if the null detector is an ideal ($R_m = 0$) 0- to 500-μA ammeter. Assume $V = 9$ V.

In our discussion of Thevenin's theorem in Appendix E we developed the Thevenin equivalent circuit for the Wheatstone bridge. Recall that

$$R_{TH} = R_1 \parallel R_2 + R_3 \parallel R_4 \tag{8-2}$$

and

$$V_{TH} = V\left(\frac{R_2}{R_1 + R_2}\right) - V\left(\frac{R_4}{R_3 + R_4}\right) \tag{8-3}$$

Therefore

$$R_{TH} = 10 \text{ kΩ} \parallel 10 \text{ kΩ} + 20 \text{ kΩ} \parallel 5 \text{ kΩ}$$
$$= 5 \text{ kΩ} + 4 \text{ kΩ}$$
$$= 9 \text{ kΩ}$$

and

$$V_{TH} = (9 \text{ V})\left(\frac{10 \text{ kΩ}}{10 \text{ kΩ} + 10 \text{ kΩ}}\right) - (9 \text{ V})\left(\frac{5 \text{ kΩ}}{20 \text{ kΩ} + 5 \text{ kΩ}}\right)$$
$$= 4.5 \text{ V} - 1.8 \text{ V}$$
$$= 2.7 \text{ V}$$

Since our ideal 0- to 500-μA ammeter has an $R_m = 0$ we can easily picture the Thevenin equivalent shown in Fig. 8-4. Thus

$$I_{AB} = \frac{V_{TH}}{R_{TH}} = \frac{2.7 \text{ V}}{9 \text{ kΩ}} = 0.3 \text{ mA} = 300 \text{ }\mu\text{A}$$

is the current in the equivalent circuit.

8-4 THE WHEATSTONE BRIDGE OHMMETER

Measurements requiring high precision and accuracy typically employ *comparison* techniques. The idea is to compare the un-

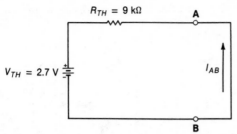

Fig. 8-4. Thevenin equivalent for Example 8-2.

known with a standard (which can be varied) until the unknown equals the standard. An example of an instrument which measures resistance by a comparison technique is the Wheatstone bridge ohmmeter. Recall that for balance the following condition must be met:

$$R_1 R_4 = R_2 R_3$$

Substituting the unknown resistance we wish to measure (R_x) for R_4 we have

$$R_1 R_x = R_2 R_3$$

$$R_x = R_2 \left(\frac{R_3}{R_1}\right)$$

The ratio R_3/R_1 is normally called the *multiplier* (M). Thus

$$R_x = M R_2 \tag{8-4}$$

In most commercial bridges M is some multiple (or fraction) of 10. An example of a practical Wheatstone bridge ohmmeter is illustrated in Fig. 8-5. An analysis of Fig. 8-5 reveals the following:

1. The multiplier (M) can assume the following values: 0.001, 0.01, 0.1, 1, 10, 100, and 1 000.
2. Resistor R_2 is a set of precision decade resistor combinations which can provide resistance values from 0.1 Ω to 9.999 9 kΩ.
3. Since $R_x = M R_2$, the Wheatstone bridge ohmmeter in Fig. 8-5 can be used to measure unknown values of resistance between 0.1 mΩ (0.001 × 0.1 Ω) and approximately 10 MΩ (1 000 × 9.999 9 kΩ).

What makes the Wheatstone bridge ohmmeter superior to the series- and shunt-type ohmmeters discussed in Chapter 5? Recall that series- and shunt-type ohmmeters have the following undesirable characteristics:

1. Nonlinear scales which are difficult to read accurately.
2. Changes in battery voltage can significantly affect the accuracy of the measurement.

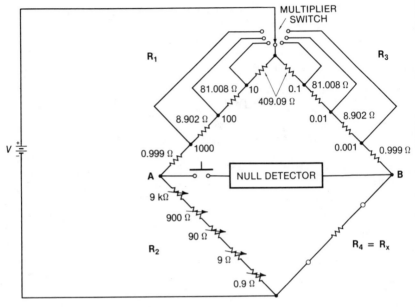

Fig. 8-5. A practical Wheatstone bridge ohmmeter.

3. Loading error.
4. Calibration error.

In Fig. 8-5 the measurement of an unknown resistor typically would be made as follows:

1. An appropriate multiplier would be selected via the multiplier switch.
2. The resistance to be measured (R_x) would be placed in the R_4 bridge arm.
3. Resistor R_2 decade would be adjusted to achieve balance. This condition would be indicated when the null detector (a sensitive ammeter) reads zero.
4. The value of R_x would be calculated from Equation 8-4:
   $R_x = MR_2$.

Note that *the measurement is made when the bridge is balanced.* Thus loading error is essentially eliminated since the load (null detector) does not draw any current when the bridge is balanced. Calibration error is not significant due to the fact that there is little ambiguity in the zero position of the null detector. Obviously, the nonlinear scales associated with the series- and shunt-type ohmmeters have been eliminated in the Wheatstone bridge ohmmeter. Finally, since balance is a function of the resistance *ratios*, changes in battery voltage will not affect the accuracy of the measurement. Thus the *principal* source of error in a Wheatstone bridge ohmmeter is a function of the tolerances of R_1, R_2, and R_3. Commercial Wheatstone bridge ohmmeters employ high-quality, precision resistors, thus ensuring very accurate measurements.

8-5 DUAL-SUPPLY BRIDGE

An example of a bridge often found in instruments and control systems is the dual-supply bridge illustrated in Fig. 8-6A. Fig. 8-6A indicates how the dual-supply bridge normally appears in schematic diagrams. Fig. 8-6B is equivalent to Fig. 8-6A, but has been redrawn to make it easier for you to follow the development of the Thevenin equivalent circuit.

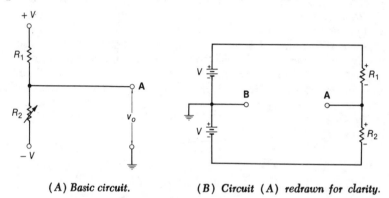

(A) *Basic circuit.*   (B) *Circuit (A) redrawn for clarity.*

Fig. 8-6. A dual-supply bridge.

Referring to Fig. 8-6B, the Thevenin equivalent resistance R_{TH} is $R_{TH} = R_{AB}$ (both voltage sources shorted). Also,

$$R_{TH} = R_1 \parallel R_2 \qquad (8\text{-}5)$$

and

$$\begin{aligned}
V_{TH} = V_{AB} &= V_{R2} - V \\
&= 2V \left(\frac{R_2}{R_1 + R_2} \right) - V \\
&= V \left(\frac{2R_2}{R_1 + R_2} - 1 \right) \\
&= V \left[\frac{2R_2 - (R_1 + R_2)}{R_1 + R_2} \right] \\
&= V \left(\frac{R_2 - R_1}{R_1 + R_2} \right) \qquad (8\text{-}6)
\end{aligned}$$

205

Equation 8-6 indicates that the dual-supply bridge will be balanced ($V_{TH} = V_{AB} = 0$) when $R_1 = R_2$.

Why are bridges important instruments for measurements? The answer to this question will be happily provided, following a brief introduction to transducers.

8-6 TRANSDUCERS

A transducer is a device or circuit that provides an electrical output (current or voltage) proportional to the input (which is *usually* a nonelectrical quantity). In effect, transducers convert nonelectrical inputs to electrical outputs. These outputs can then be measured, processed, and/or controlled electronically. There is almost an endless variety of transducers. In this section we will discuss a few of the most common types and see how they "fit in" with the bridge circuits discussed previously.

From your study of electricity you know that the resistance of a conductor depends upon the following factors: length, diameter, type of material, and temperature. If the conductor is in an environment where the temperature is constant, or doesn't vary significantly, then (for a given material) resistance is a function of the length and diameter of the conductor. This fact is summarized by the following equation:

$$R = \rho \frac{l}{A} \qquad (8\text{-}7)$$

where
R = resistance of the conductor,
ρ = resistivity, which is a material constant,
l = length of the conductor,
A = cross-sectional area of the conductor.

Strain Gages

A *strain gage* is a transducer that takes advantage of the relationship described by Equation 8-7 in order to measure strain. Strain refers to the change in the dimensions of an object subjected to a force that tends to compress or stretch the object. Specifically, strain (S) is defined as

$$S = \frac{\Delta L}{L} \qquad (8\text{-}8)$$

where
ΔL = change in length,
L = original length.

206

EXAMPLE 8-3

A bar 10 cm long is stretched until it is 10.2 cm long. Determine the strain.

From Equation 8-8,

$$S = \frac{\Delta L}{L} = \frac{10.2 \text{ cm} - 10 \text{ cm}}{10 \text{ cm}} = 0.02 = 2\%$$

is the strain.

In its most basic form a strain gage is nothing more than a thin piece of wire attached to a small insulating material. This is illustrated in Fig. 8-7. If the strain gage is bonded to a test specimen which is then stretched (or compressed) its resistance will change,

Fig. 8-7. A simple strain gage.

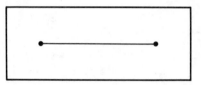

because the length and diameter of the strain gage are changed when the specimen is stretched (or compressed). The relationship between the change in resistance of the strain gage and the change in length of the specimen is

$$\frac{\Delta R}{R} = K \frac{\Delta L}{L} \tag{8-9}$$

The term K is a constant called the *gage factor* and is specified by the manufacturer of the strain gage. Substituting S for $\Delta L/L$ in Equation 8-9 yields

$$\frac{\Delta R}{R} = KS \tag{8-10}$$

Figs. 8-8 and 8-9 illustrate how strain can be measured utilizing a strain gage *and* the bridges discussed previously.

In Fig. 8-8 we could select values for R_1, R_2, and R_3 so that $R_1 = R_2 = R_3 = R$, where R is the initial (unstrained) resistance of the strain gage. Thus the bridge is initially balanced so that the null detector reads zero. The test specimen would then be stretched (or compressed)—this would change the resistance of the strain gage and thus the bridge would become unbalanced. This condition would be indicated by deflection of the null detector. If the bridge is now balanced by adjustment of R_3, the value of R_3 is proportional to the

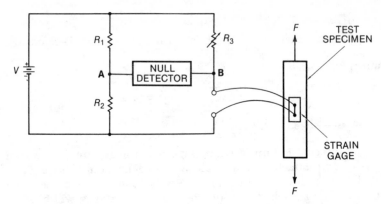

Fig. 8-8. Measuring strain with a Wheatstone bridge.

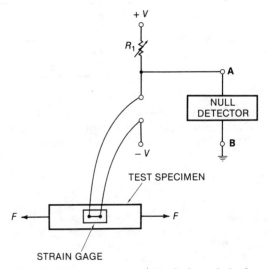

Fig. 8-9. Measuring strain with a dual-supply bridge.

strain; therefore R_3 could be calibrated to read the strain directly. A similar procedure would be followed for the bridge in Fig. 8-9.

EXAMPLE 8-4

In Fig. 8-8 let $R_1 = R_2 = 1\ 500\ \Omega$. Resistor R_3 is a potentiometer which is initially adjusted to $1\ 500\ \Omega$, and R_4 is a strain gage which has been bonded to a steel test specimen. The strain gage has a resistance (unstrained) of $1\ 500\ \Omega$ and a gage factor (K) of 1.5. After the steel test specimen is stretched it is necessary to adjust R_3 to $1\ 504.5\ \Omega$ in order to balance the bridge. Determine:

(a) The percent of strain.

(*b*) The final length of the test specimen, assuming the initial length was 5.08 cm.

(*a*)

$$\frac{\Delta R}{R} = KS$$

Therefore

$$S = \frac{\Delta R}{KR} = \frac{1\,504.5\,\Omega - 1\,500\,\Omega}{1.5(1\,500\,\Omega)}$$
$$= 0.002 = 0.2\%$$

(*b*)

$$\frac{\Delta L}{L} = S$$
$$\Delta L = LS$$
$$= (5.08\text{ cm})(0.002)$$
$$= 0.010\,16\text{ cm}$$

The final length is then
$$L_f = L + \Delta L$$
$$= 5.08\text{ cm} + 0.010\,16\text{ cm}$$
$$\cong 5.09\text{ cm}$$

In practice, the measurement of strain is often more complicated than the procedure illustrated in Example 8-4. Typical problems include the following:

1. Selecting an appropriate *type* of strain gage from a wide variety of commercially available gages.
2. Determining the *number* of gages required for the structure to be tested.
3. Determining *where* the gages should be mounted on the structure to be tested.
4. Selecting a suitable *bonding technique* that is compatible with the gage and the structure to be tested.
5. Ensuring that the bridge is *initially* balanced. This is necessary because the resistance (unstrained) of strain gages vary slightly from one to another.

In spite of the problems just cited (and others), strain gages are widely used in industry. Readers wishing to pursue this topic further are referred to manufacturers' catalogs and texts devoted to strain gage measurements.

Position Transducers

Potentiometers are frequently employed as position transducers. In such applications a voltage is produced that is analogous to posi-

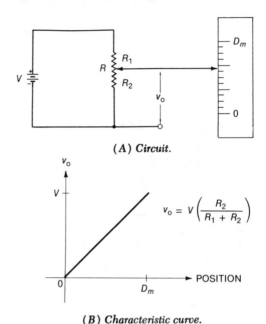

(A) Circuit.

$$v_o = V\left(\frac{R_2}{R_1 + R_2}\right)$$

(B) Characteristic curve.

Fig. 8-10. A simple position transducer.

tion. This concept is illustrated in Fig. 8-10. Note in Fig. 8-10A that the potentiometer wiper is mechanically ganged to the shaft of the position indicator. Thus, as the position indicator travels along the calibrated scale, the voltage across R_2 varies proportionally. The relationship between v_o and position is illustrated graphically in Fig. 8-10B. Slide-type potentiometers are used to detect rectilinear motion, while rotary potentiometers are employed to detect angular motion. An example of the latter case is illustrated in Fig. 8-11A, where the shaft of a rotary-type potentiometer has been coupled to the shaft of a motor. Fig. 8-11B illustrates the details of the potentiometer circuit. Notice in Fig. 8-11B that as the shaft of the motor rotates, the voltage across R_2 (v_o) varies proportionally. Thus the output voltage is an indication of the angular position of the motor shaft. Fig. 8-11C graphically illustrates the relationship between output voltage and shaft position.

By utilizing a dual-supply bridge and slide-type potentiometer you can construct a position transducer like the one illustrated in Fig. 8-12. Such a circuit could be used to accurately align two parts in a manufacturing process. There are, of course, many other types of position transducers available, some of which are quite complex. Yet the simple potentiometer is one of the most common position transducers you are likely to encounter!

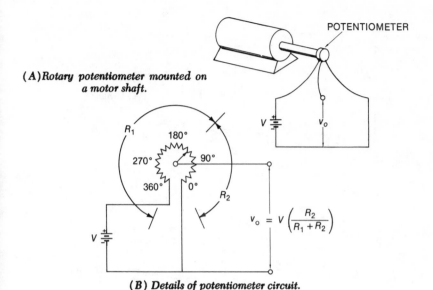

(A) Rotary potentiometer mounted on a motor shaft.

$$v_o = V \left(\frac{R_2}{R_1 + R_2} \right)$$

(B) Details of potentiometer circuit.

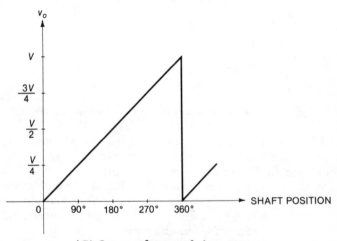

(C) Output voltage vs. shaft position.

Fig. 8-11. Detecting angular motion.

For virtually every nonelectrical quantity you wish to measure you can probably find a suitable transducer.

8-7 ELECTRONIC INSTRUMENTS

From your study of electronic devices and circuits perhaps you have acquired a working knowledge of operational amplifiers (op

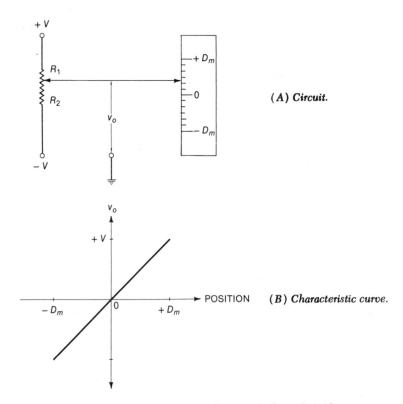

Fig. 8-12. Position transducer employing a dual-supply bridge.

amps).* We do not have the time to explore the fascinating topic of op amps in great detail. We will, however, examine several op-amp circuits apropos to our study of analog instrumentation. You can think of these circuits as *electronic building blocks* that will enable you to construct simple, yet very practical, electronic instruments. As you will soon see, these building blocks "fit in" nicely with the instruments you are already familiar with.

Op amps have the following very desirable characteristics: large gain, high input resistance, low output resistance, and low cost. In addition they are easy to design with! One of the few disadvantages of op amps is that they normally require two power supplies. The symbol for an op amp is illustrated in Fig. 8-13A (for simplicity the power supplies are not shown). Note that the op amp has two inputs: inverting and noninverting. The circuits we will discuss (with

---

* If this is not the case, the two references at the end of the chapter are excellent texts to consult.

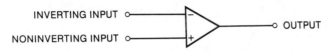

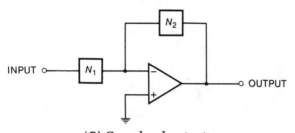

(A) *Op-amp symbol.*

(B) *General analog circuit.*

Fig. 8-13. The op amp.

one exception) fit the general model illustrated in Fig. 8-13B. Notice in Fig. 8-13B that the noninverting input (+) has been grounded. The blocks labeled N_1 and N_2 in Fig. 8-13B are passive networks, i.e., combinations of resistors, capacitors, and/or inductors. By selecting appropriate networks for N_1 and N_2 we obtain the circuits illustrated in Fig. 8-14. These circuits perform the following functions:

a. *Differentiator*—Provides an output voltage proportional to the rate of change (slope) of the input voltage.

$$v_o = -RC \frac{dv}{dt} \tag{8-11}$$

The symbol $\frac{dv}{dt}$ represents the *slope* of the input voltage.

b. *I/V Transducer*—Provides an output voltage proportional to an input current.

$$v_o = IR \tag{8-12}$$

c. *Voltage Amplifier*—Provides an output voltage larger than the input voltage. Notice in Fig. 8-14C that the noninverting (+) input lead of the op amp is where the input voltage is applied to the circuit. This is the exception to the general model shown in Fig. 8-13B.

$$v_o = v_{in} \left(\frac{R_1}{R_2} + 1 \right) \tag{8-13}$$

Suppose it is necessary to obtain information about the velocity and acceleration of some object. With a position transducer you can

213

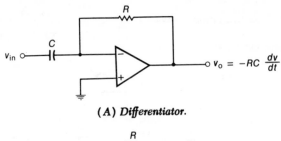

(A) Differentiator.

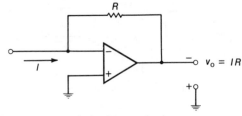

(B) I/V transducer.

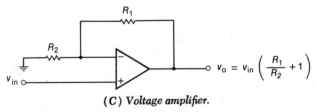

(C) Voltage amplifier.

Fig. 8-14. Basic instrumentation op-amp circuits.

generate an analog signal proportional to the object's position (x). The relationships between position (x), velocity (v), and acceleration (a) are as follows:

$$v = \frac{dx}{dt}$$

and

$$a = \frac{dv}{dt}$$

In words, the velocity (v) of an object is the rate of change (slope) of its position (x) with respect to time (t), and the acceleration (a) of an object is the rate of change (slope) of its velocity (v) with respect to time (t). Thus by applying the analog position signal to the input of a differentiator (Fig. 8-14A) you obtain an output signal proportional to velocity. Similarly, the analog velocity signal could be applied to the input of a second differentiator to obtain a signal proportional to acceleration. The same information could be acquired with velocity and acceleration transducers; such transducers tend to be expensive, however. By employing a relatively simple position transducer and low-cost op-amp differentiators, it is often possible to generate the desired information economically. This concept is illustrated in Fig. 8-15. The output voltages in Fig. 8-15 could drive a chart recorder, oscilloscope, etc., in order to obtain a graphical representation of the responses.

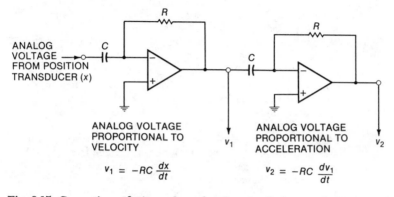

Fig. 8-15. Generating velocity and acceleration signals from a position signal.

Recall from our discussion of rectifier- and peak-type ac voltmeters, that, for small ac signals, average diode current is *not* directly proportional to the amplitude of the input voltage. This is because a small input voltage establishes the diode's operating point in the nonlinear portion of the diode's I/V curve. Thus many vom's have a separate nonlinear scale for small ac voltage measurements. One solution to this problem is to amplify small ac signals *prior* to measuring them with rectifier- or peak-type voltmeters. This concept is illustrated in Fig. 8-16, where the circuit of Fig. 8-14C is employed for the amplifier. In Fig. 8-16 if $R_1 = 9R_2$ then the gain of the amplifier from Equation 8-13 would be 10. Thus a 0.7-V input voltage to the amplifier would produce a 7-V output voltage to drive the recti-

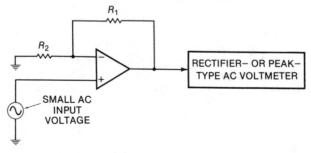

Fig. 8-16. One solution to the problem of measuring small ac voltages.

fier- or peak-type ac voltmeter. This would permit the use of a linear scale for small ac voltages. The following examples illustrate how the other op-amp circuits in Fig. 8-14 can be employed to measure current, voltage, and resistance.

EXAMPLE 8-5

(a) Design an electronic ammeter which employs the I/V transducer of Fig. 8-14B. The ammeter should have a range of 0–10 μA.

(b) Compare the characteristics of the electronic ammeter with a standard analog meter movement.

(a) The idea is to employ the I/V transducer (often called an I/V converter) to generate a voltage proportional to the current you wish to measure. The output voltage from the I/V transducer is then measured with a dc voltmeter which has been calibrated to read current. We will assume our 50-μA, 5-kΩ meter movement is available for the dc voltmeter portion of the instrument. The maximum output voltage from the I/V transducer is limited by the op-amp's supply voltages. If we pick supply voltages of ± 15 V for the op amp, limiting the output voltage to 10 V would be reasonable. This choice also has the advantage of providing a convenient ratio of output voltage to input current (1 volt per microampere). Thus:

Calculate R for the op amp:

$$v_o = IR$$

or

$$R = \frac{v_o}{I} = \frac{v_{o(max)}}{I_{max}} = \frac{10\ V}{10\ \mu A} = 1\ M\Omega$$

Calculate R_s for the dc voltmeter:

$$R_{in} = SV_{FS} = \left(20\ \frac{k\Omega}{V}\right)(10\ V)$$
$$= 200\ k\Omega$$
$$R_s = R_{in} - R_m$$
$$= 200\ k\Omega - 5\ k\Omega = 195\ k\Omega$$

The design employing a 741 op amp is illustrated in Fig. 8-17.

(b) The specifications for a *typical* 741-type op amp indicate that it has an open-loop gain (no feedback) of 200 000. The input resistance of an I/V transducer like the one employed is (approximately) given by the following formula:

$$R_{in} = \frac{R}{A_{OL}} \qquad (8\text{-}14)$$

Thus

$$R_{in} = \frac{1\ M\Omega}{200\ 000} = 5\ \Omega!$$

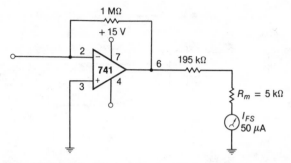

Fig. 8-17. A 0- to 10-μA electronic ammeter (Example 8-5).

This is a very small input resistance compared to the 5-kΩ of the 50-μA meter movement! Thus the loading effects of the electronic ammeter will also be very small compared to a standard analog meter movement! This is a very significant advantage if you must measure small currents in low-resistance circuits. By employing a better-quality op amp it is possible to construct a nanoampere ammeter which also has a very small input resistance. By employing op amps and analog instruments you can design practical, inexpensive, high-quality instruments useful for a variety of measurements.

EXAMPLE 8-6

Design an electronic ohmmeter which employs a 1-mA current source and the I/V transducer of Fig. 8-14B.

The I/V transducer converts an input current to an output voltage. If the input current is constant, then the output voltage will be proportional to the resistance R in Fig. 8-14B. Assuming that the input current is fixed at 1 mA, the output voltage will be 10 V when $R = 10$ kΩ, 9 V when $R = 9$ kΩ, 8 V when $R = 8$ kΩ, etc. To design a simple electronic ohmmeter you can employ the following procedure.

1. Define the desired resistance range.
2. Select supply voltages for the op amp.
3. Specify $v_{o(max)}$. As a rule of thumb limit $v_{o(max)}$ from two-thirds to three-fourths of the supply voltages.
4. Calculate the required value of the current source by

$$I = \frac{v_{o(max)}}{R_{max}} \qquad (8\text{-}15)$$

To illustrate the procedure:
1. Range: 0–10 kΩ.
2. Select ± 15 V for supply voltages.
3. Limit $v_{o(max)}$ to two-thirds of 15 V, or 10 V.
4. $I = \dfrac{v_{o(max)}}{R_{max}} = \dfrac{10 \text{ V}}{10 \text{ k}\Omega} = 1$ mA

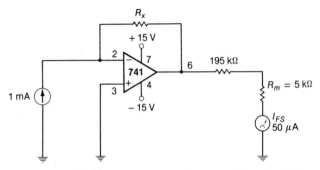

Fig. 8-18. A 0- to 10-kΩ electronic ohmmeter (Example 8-6).

Our design is illustrated in Fig. 8-18.

EXAMPLE 8-7

Design an electronic voltmeter employing the voltage amplifier of Fig. 8-14C. The voltmeter should provide a 0- to 100-mV range and a 0- to 1-V range.

You can employ the following procedure for the design:

1. Specify the desired voltage ranges.
2. Select supply voltages for the op amp.
3. Specify $v_{o(\max)}$ (two-thirds to three-fourths rule).
4. Calculate the required voltage gain of the voltage amplifier (A). This is determined by dividing $v_{o(\max)}$ by the maximum input voltage of the *lowest* range.
5. Determine the ratio of R_1 to R_2. Solving Equation 8-13 for R_1/R_2 yields

$$\frac{R_1}{R_2} = A - 1$$

6. Select values for R_1 and R_2.
7. Design a voltage divider for the input to the voltmeter.

Thus:

1. The desired ranges are 0–100 mV and 0–1 V.
2. Select ±15 V for supply voltages.
3. Limit $v_{o(\max)}$ to two-thirds of 15 V, or 10 V.
4. $A = \dfrac{10\text{ V}}{100\text{ mV}} = 100$
5. $\dfrac{R_1}{R_2} = A - 1 = 100 - 1 = 99$
6. Pick $R_1 = 99$ kΩ and $R_2 = 1$ kΩ.
7. The output voltage of the op amp should be 10 V when the input to the op amp is 100 mV. Thus, when 1 V is applied to the input of the voltage divider we want the output to be 100 mV. This requires a 9:1 ratio for the voltage divider resistors. Fig. 8-19 illustrates the completed design. In order to keep the input resistance of the circuit high reasonably large resistances are normally employed for the voltage divider.

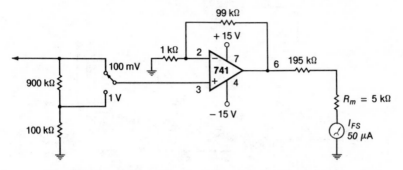

Fig. 8-19. A multiple-range electronic voltmeter (Example 8-7).

Fig. 8-20 illustrates a *difference* amplifier. *As the name implies, a difference amplifier provides an output voltage proportional to the difference between two input voltages.* In instrumentation systems the inputs of the difference amplifier are often wired to the output of a bridge circuit containing one or more transducers. For this reason

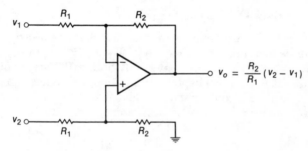

Fig. 8-20. A balanced difference amplifier.

difference amplifiers are often referred to as *bridge* or *transducer* amplifiers. Fig. 8-21 is an example of a typical instrumentation amplifier, employing three op amps. Note in Fig. 8-21 that the third op amp and associated resistors constitute the difference amplifier illustrated in Fig. 8-20. The op amps labeled 1 and 2 in Fig. 8-21 are a "special case" of the voltage amplifier illustrated in Fig. 8-14C, where $R_1 = 0$ and $R_2 = \infty$. If you substitute 0 for R_1 and ∞ for R_2 into Equation 8-13 you can see that the *resulting voltage gain is unity* ($v_o = v_{in}$). Perhaps you are wondering why we would want an amplifier whose voltage gain is unity. For the "special case" just described the input resistance of the op amp circuit will be *very* large. This type of op-amp circuit is called a *voltage follower*. Voltage followers are often used as *buffers* between sources and low-resistance loads to reduce loading effects. There are many other op-amp circuits

219

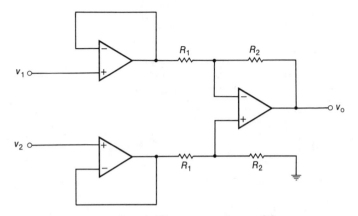

Fig. 8-21. A typical instrumentation amplifier.

that are used in instrumentation applications. The interested reader is referred to texts dealing with op-amp applications.

8-8 REVIEW OF OBJECTIVES

Dc bridges are used with transducers so that nonelectrical quantities can be measured electronically. Two popular dc bridges are the Wheatstone bridge and the dual-supply bridge. Both types of bridges were discussed in this chapter. In addition, the Wheatstone bridge is often used as a comparison instrument to accurately measure resistance. Such an instrument has minimal calibration and loading error.

Transducers provide electrical outputs proportional to the inputs, which are usually (but not always) nonelectrical in nature. Examples were provided illustrating the measurement of strain, position, velocity, and acceleration. By combining dc bridges, transducers, and op-amp circuits, it is possible to construct electronic instruments. Examples of elementary electronic ammeters, voltmeters, and ohmmeters were provided. Readers are encouraged to expand their knowledge of op-amp circuits by consulting texts dealing specifically with these fascinating devices.

8-9 QUESTIONS

1. When will a Wheatstone bridge be balanced?
2. When will a dual-supply bridge be balanced?
3. Why is a Wheatstone bridge ohmmeter more accurate than a series- or shunt-type ohmmeter?
4. What is the principal factor that determines the accuracy of a Wheatstone bridge ohmmeter?

5. What is a transducer?

6. Define strain, and explain how it can be measured.

7. What is a differentiator?

8. Explain how a potentiometer can be used as a position indicator.

9. Why are op amps important?

10. What advantages do electronic instruments have over standard analog instruments?

11. Dr. Courtine recommends using a 0- to 1-A ammeter as a null detector for a Wheatstone bridge ohmmeter. Explain why this is a poor choice.

8-10 PROBLEMS

1. (a) What value of R_4 is required to balance the bridge in Fig. 8-22A?
   (b) Assume $R_4 = 15$ kΩ. Determine V_{AB} and I_{AB} if the ammeter is a 10-μA, 100-Ω meter movement.

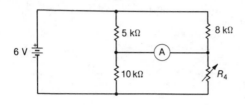

(A) Problems 1 and 2.

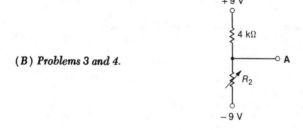

(B) Problems 3 and 4.

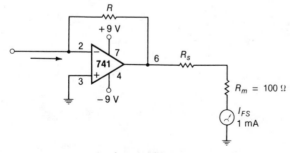

(C) Problems 8 and 9.

Fig. 8-22. Schematic diagrams for Chapter 8 problems.

2. Repeat Problem 1(b) for $R_4 = 17$ kΩ.

3. (a) What value of R_2 is required to balance the bridge in Fig. 8-22B?
   (b) Determine V_A if $R_2 = 3.9$ kΩ.
   (c) Determine V_A if $R_2 = 4.1$ kΩ.

4. Resistor R_2 in Fig. 8-22B is a *thermistor*. A thermistor's resistance varies with temperature. We will assume the thermistor is described by the following relationship:

$$R = R_{25}(1 + \alpha \Delta t) \qquad (8\text{-}16)$$

where

R = resistance at some temperature t,
R_{25} = resistance at 25°C,
α = thermistor's temperature coefficient,
$\Delta t = t - 25$°C.

Assume the thermistor has a resistance of 4 kΩ at 25°C and that $\alpha = -3.33 \times 10^{-4}$ Ω/°C. Estimate the output voltage (V_A) if the temperature increases from 25°C to 100°C.

5. The output voltage in Fig. 8-11B is 6.75 V. Assuming $V = 9$ V, estimate the position of the shaft in degrees.

6. The circuit in Fig. 8-10 employs a 19-kΩ potentiometer, and 9-V battery. The scale is calibrated in 1-mm intervals and extends from 0 to 12 mm. Assuming $V_o = 2.25$ V, estimate:
   (a) R_1 and R_2.
   (b) The position of the pointer on the calibrated scale.

7. The Wheatstone bridge in Fig. 8-5 is used to measure a resistance of 8 419.9 Ω. What values of R_1, R_2, and R_3 are required for balance? Hint: assume $M = 1$.

8. (a) Determine the values of R and R_s in Fig. 8-22C so that currents between 0 and 100 μA can be measured. Assume a 5-V output should result when $I = 50$ μA.
   (b) Estimate the input resistance of the electronic ammeter.

9. Modify the circuit in Fig. 8-22C so that resistances between 0 and 25 kΩ can be measured.

10. Design an electronic voltmeter that provides a 0- to 250-mV range and a 0- to 2-V range. Assume that two 9-V batteries and a 1-mA, 100-Ω meter movement are available for the design.

8-11   EXPERIMENT 8-1

Objective

The objective of this experiment is to design an elementary Wheatstone bridge ohmmeter.

Material Required

9-V battery
Resistor decades, variable in 100-kΩ, 10-kΩ, 1-kΩ, 100-Ω, and 10-Ω steps
50-μA meter movement

Introduction

A Wheatstone bridge will be balanced when the cross products of the resistances are equal. Specifically, balance occurs when

$$R_1 R_4 = R_2 R_3 \qquad (8\text{-}1)$$

The unknown resistance we wish to measure (R_x) is placed in the R_4 bridge arm. Thus

$$R_1 R_x = R_2 R_3$$

$$R_x = \frac{R_2 R_3}{R_1} = M R_2 \qquad (8\text{-}4)$$

The ratio $M = R_3/R_1$ is called the *multiplier* since it determines how much larger than R_x resistor R_2 must be for balance. The method we will employ to design our Wheatstone bridge ohmmeter is as follows:

1. Specify the range of unknown resistances you wish to measure.
2. Select a multiplier (M).
3. Select values for R_1 and R_3.
4. Determine the required range of R_2, so that it is possible to balance the bridge for all values of R_x specified in Step 1. The maximum value of R_2 required to balance the bridge when R_x is maximum is

$$R_{2(\text{max})} = \frac{R_1 \, R_{x(\text{max})}}{R_3} = \frac{R_{x(\text{max})}}{M} \qquad (8\text{-}17)$$

As an example we will illustrate the design of a 0- to 1-MΩ Wheatstone bridge ohmmeter.

1. The specified range in R_x is from 0 to 1 MΩ.
2. Select $M = 10$.
3. Pick $R_3 = 10 \text{ k}\Omega$ and $R_1 = 1 \text{ k}\Omega$. This choice ensures that $M = 10$.
4. $R_{2(\text{max})} = \dfrac{R_{x(\text{max})}}{M} = \dfrac{1 \text{ M}\Omega}{10} = 100 \text{ k}\Omega$

With the addition of a 9-V battery and voltmeter for the null detector our design is illustrated in Fig. 8-23.

Procedure

Step 1. Design a 0- to 10-V dc voltmeter which employs your 50-μA meter movement. You will use this voltmeter as a null detector.

Step 2. Build the Wheatstone bridge ohmmeter illustrated in Fig. 8-23.

Step 3. For each of the R_x values 1 MΩ, 100 kΩ, 10 kΩ, 1 kΩ, 100 Ω,

223

and 10 Ω, adjust R_2 until balance is achieved. In each case record the appropriate value of R_2.

$R_x = 1\,M\Omega, R_2 = $ _____ $R_x = 1\,k\Omega, R_2 = $ _____
$R_x = 100\,k\Omega, R_2 = $ _____ $R_x = 100\,\Omega, R_2 = $ _____
$R_x = 10\,k\Omega, R_2 = $ _____ $R_x = 10\,\Omega, R_2 = $ _____

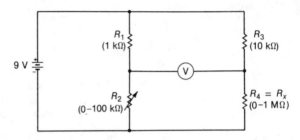

Fig. 8-23. A 0- to 1-MΩ Wheatstone bridge ohmmeter for Experiment 8-1.

Step 4. Calculate the "measured value" of R_x from Equation 8-4:
$R_x = MR_2$.

Step 5. Redesign the Wheatstone bridge ohmmeter so that resistances between 0 and 100 kΩ can be measured. In this case the multiplier M should be 2.

Step 6. Verify your design by measuring five resistors as follows: 4.7 kΩ, 10 kΩ, 22 kΩ, 68 kΩ, and 100 kΩ. How do your measured values compare with the color coded values?

Conclusion

What limited the accuracy of the Wheatstone bridge ohmmeters employed in this experiment? How would you modify the bridges to provide several resistance ranges?

8-12 EXPERIMENT 8-2

Objective

The objective of this experiment is to investigate the characteristics of elementary electronic instruments.

Material Required

50-μA meter movement
$\pm$15-V power supplies
5-V power supply
741 op amp
Resistor decades, variable in 100-kΩ, 10-kΩ, 1-kΩ, and 100-Ω steps

224

Introduction

Fig. 8-24A illustrates an op-amp current-to-voltage (I/V) transducer. Since the output voltage (v_o) equals IR, you can measure v_o with a dc voltmeter which has been calibrated to indicate the current I. The steps in the design of an electronic ammeter like the one illustrated in Fig. 8-24B are as follows.

1. Specify the desired current range.
2. Select supply voltages for the op amp.
3. Specify $v_{o(max)}$. Limit $v_{o(max)}$ from two-thirds to three-fourths of the supply voltage.
4. Calculate the required value of R from Ohm's law:

$$R = \frac{V_{o(max)}}{I_{max}}$$

If the input current to an I/V transducer is *constant*, then the output voltage depends upon the value of R. This fact permits us to measure resistance with a circuit like the electronic ohmmeter illustrated in Fig. 8-24C. Naturally the voltmeter in Fig. 8-24C is calibrated to indicate resistance rather than voltage. The following steps can be used to design the electronic ohmmeter.

1. Specify the desired resistance range.
2. Select supply voltages for the op amp.
3. Specify $v_{o(max)}$. Limit $v_{o(max)}$ from two-thirds to three-fourths of the supply voltage.
4. Calculate the required value of the current source:

$$I = \frac{v_{o(max)}}{R_{max}}$$

In this experiment we will use a 741 op amp since these op amps are inexpensive and easily acquired.

Procedure

Step 1. Calculate the value of R_s in Fig. 8-24B to convert the 50-μA meter movement into a 0- to 10-V dc voltmeter.
$R_s = \underline{\hspace{3cm}}$

Step 2. Design the electronic ammeter in Fig. 8-24B so that currents between 0 and 20 μA can be measured. Note that the supply voltages for the op amp are ±15 V, and that $v_{o(max)}$ will be limited to 10 V.

Step 3. Estimate the input resistance of your electronic ammeter from $R_{in} = R/A_{OL}$. You can assume A_{OL} is 200 000 for the 741 op amp.
$R_{in} = \underline{\hspace{3cm}}$

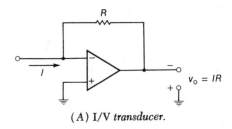

(A) I/V transducer.

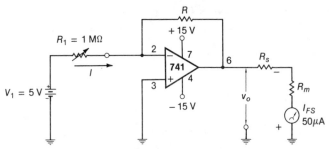

(B) Electronic ammeter.

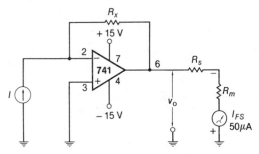

(C) Electronic ohmmeter.

Fig. 8-24. Schematic diagrams for Experiment 8-2.

Step 4. Since the value of R_{in} calculated in Step 3 is small compared with R_1, the input current (I) in Fig. 8-24B is closely approximated by V_1/R_1. Calculate the value of I for the following values of R_1: 1 MΩ, 500 kΩ, 333.3 kΩ, and 250 kΩ.
For $R_1 = 1$ MΩ, $I =$ _____   For $R_1 = 500$ kΩ, $I =$ _____
For $R_1 = 333.3$ kΩ, $I =$ _____   For $R_1 = 250$ kΩ, $I =$ _____

Step 5. Build the electronic ammeter. Record the measured currents when R_1 is 1 MΩ, 500 kΩ, 333.3 kΩ, and 250 kΩ. Compare measured and calculated values of I.
For $R_1 = 1$ MΩ, $I =$ _____   For $R_1 = 500$ kΩ, $I =$ _____
For $R_1 = 333.3$ kΩ, $I =$ _____   For $R_1 = 250$ kΩ, $I =$ _____

Step 6. Design the electronic ohmmeter illustrated in Fig. 8-24C,

226

so that resistances between 0 and 100 kΩ can be measured. Again we will use ±15 V for the op-amp supply voltages, and limit the maximum output voltage ($v_{o(max)}$) to 10 V.

Step 7. We will employ the 5-V source (V_1) and R_1 to provide the required constant current source illustrated in Fig. 8-24C. Adjust R_1 to provide the required current I.

Step 8. Vary R_x in Fig. 8-24C in 10-kΩ steps from 10 kΩ to 100 kΩ. In each case record the "measured value" of R_x as indicated by the amount of deflection (D) of the voltmeter. (A 100-kΩ resistor should produce full-scale deflection, 50-kΩ half-scale deflection, etc.)

$R_x = 10\ \text{kΩ}, D = \underline{\hspace{2cm}}$ $R_x = 60\ \text{kΩ}, D = \underline{\hspace{2cm}}$
$R_x = 20\ \text{kΩ}, D = \underline{\hspace{2cm}}$ $R_x = 70\ \text{kΩ}, D = \underline{\hspace{2cm}}$
$R_x = 30\ \text{kΩ}, D = \underline{\hspace{2cm}}$ $R_x = 80\ \text{kΩ}, D = \underline{\hspace{2cm}}$
$R_x = 40\ \text{kΩ}, D = \underline{\hspace{2cm}}$ $R_x = 90\ \text{kΩ}, D = \underline{\hspace{2cm}}$
$R_x = 50\ \text{kΩ}, D = \underline{\hspace{2cm}}$ $R_x = 100\ \text{kΩ}, D = \underline{\hspace{2cm}}$

Conclusion

How would you modify the circuits employed in this experiment to obtain multiple ranges? What advantages do the instruments built in this experiment have over standard analog instruments?

8-13   REFERENCES

1. Berlin, Howard M. 1977. *Design of Op-Amp Circuits, With Experiments,* Indianapolis: Howard W. Sams & Co., Inc.
2. Berlin, Howard M. 1977. *Design of Active Filters, With Experiments,* Indianapolis: Howard W. Sams & Co., Inc.

CHAPTER 9

Passive RC Filters

9-1 INTRODUCTION

Filters are important building blocks in many instruments and instrumentation systems. A filter is a network or circuit designed to pass certain frequencies and reject other frequencies. Passive filters employ passive components such as resistors, capacitors, and/or inductors to provide the desired filtering action. Active filters combine passive components with active components, such as op amps, to improve the characteristics of the filter. An understanding of passive filters is necessary to design active filters. After reading this chapter if you would like to know more about filters the author would recommend acquiring Howard Berlin's *Design of Active Filters, With Experiments*, which is listed in Section 8-13.

9-2 OBJECTIVES

At the end of this chapter you will be able to do the following:

- Explain the difference between high-pass, low-pass, bandstop, and bandpass filters.
- Design passive *RC* high-pass and low-pass filters.
- Design a Wien-bridge bandstop filter.
- Design an elementary bandpass filter.
- Sketch Bode diagrams for filters.
- Discuss filter loading effects.

9-3 IDEAL FILTERS

There are four basic types of filters. Before we discuss them some definitions are in order:

passband—The range of frequencies that can pass through the filter.

cutoff frequency (f_c)—A frequency above or below which signals are rejected by the filter.

gain (K)—A ratio of filter output voltage to filter input voltage.

attenuation (a)—A ratio of filter input voltage to filter output voltage. Attenuation and gain are reciprocals. Thus

$$a = \frac{1}{K} \qquad (9\text{-}1)$$

decibel gain and attenuation—Tradition dictates that we employ logarithmic ratios for gain and attenuation. Specifically:

For a current or voltage ratio:

$$K_{dB} = 20 \log K \qquad (9\text{-}2)$$

$$a_{dB} = 20 \log a \qquad (9\text{-}3)$$

For a power ratio:

$$P_{dB} = 10 \log P \qquad (9\text{-}4)$$

where P is normally the ratio of output power to input power.

Gain vs. frequency curves for the four basic filter types are shown in Fig. 9-1 assuming *ideal* filters. The shaded portions in Fig. 9-1 indicate those frequencies which the filter passes. Similarly, the unshaded portions indicate those frequencies that the filter rejects. As you will see, the name for each filter describes the specific filtering characteristics of that type of filter.

a. *Low-pass filter*—Passes low frequencies, blocks (rejects) high frequencies. Signals whose frequencies are between 0 Hz (dc) and the cutoff frequency (f_c) are passed. Frequencies higher than f_c are blocked.

b. *High-pass filter*—Passes high frequencies, blocks (rejects) low frequencies. Signals whose frequencies are higher than the cutoff frequency (f_c) are passed. Frequencies lower than f_c are blocked.

c. *Bandpass filter*—This type of filter has two cutoff frequencies— a lower cutoff frequency (f_1) and an upper cutoff frequency (f_2). Signals whose frequencies are between f_1 and f_2 are passed. All other frequencies are blocked.

d. *Bandstop filter*—This type of filter also has a lower (f_1) and upper (f_2) cutoff frequency. In this case, however, signals whose frequencies are between f_1 and f_2 are blocked. Frequencies between 0 Hz (dc) and f_1 and frequencies higher than f_2 are passed.

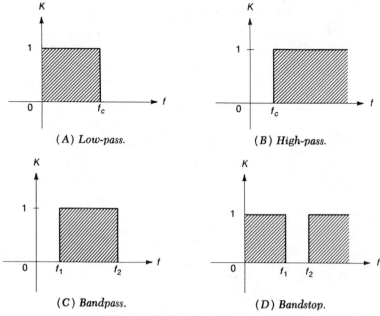

(A) Low-pass. (B) High-pass.

(C) Bandpass. (D) Bandstop.

Fig. 9-1. Ideal filter characteristics.

What makes the filters in Fig. 9-1 ideal? We can point to two outstanding features:

1. The *passband gain* (K) and attenuation (a) are *unity* ($K = a = 1$).
2. Signals whose frequencies are outside the passband are *completely* rejected ($K = 0, a = \infty$).

9-4 THE RC LOW-PASS FILTER

An RC low-pass filter and frequency response curve are shown in Fig. 9-2. Note that the actual response (solid line) differs from the ideal response (dashed line) in two important respects:

1. The passband gain does *not* remain at unity but decreases as frequency increases.
2. Signals whose frequencies are greater than f_c are attenuated but *not* completely rejected. Above f_c the gain "rolls off" gradually rather than abruptly dropping to zero.

Thus the RC low-pass filter is a crude (but useful) approximation of an ideal low-pass filter. Note in Fig. 9-2B that the cutoff frequency (f_c) is defined as the frequency where the gain (K) has dropped to 0.707.

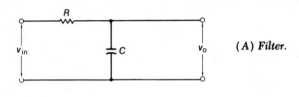

(A) *Filter.*

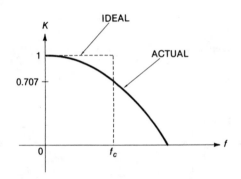

(B) *Gain vs. frequency.*

Fig. 9-2. A low-pass RC filter.

How does the low-pass filter in Fig. 9-2A work? For any frequency the input voltage will divide between the resistor (R) and capacitor (C). As you know, capacitive reactance (X_C) decreases as frequency increases. Thus for very low frequencies the capacitive reactance X_C will be very large and will approximate an open circuit. In such a case the output voltage value will be very close to the input voltage value. At high frequencies the capacitive reactance will be very small and thus approximate a short circuit. Thus for high frequencies the output voltage will be much smaller than the input voltage. For frequencies between very low and very high frequencies the input voltage will divide in such a manner that the output voltage is somewhere between 0 V and v_{in}. At the cutoff frequency (f_c) the capacitive reactance *equals* the value of resistance R. This fact is sufficient for us to derive an expression for the cutoff frequency (f_c) in terms of R and C. At f_c,

$$|X_C| = R$$

Thus

$$X_C = \frac{1}{2\pi f_c C} = R$$

or

$$2\pi f_c CR = 1$$

or

$$f_c = \frac{1}{2\pi RC} \tag{9-5}$$

where R and C are as shown in Fig. 9-2A.

In order to provide more details about the performance of the RC low-pass filter it is necessary to derive an equation for gain (K) in terms of R, C, and frequency. This derivation assumes a knowledge on the part of the reader of complex (not complicated) numbers and phasors. Readers not familiar with these concepts should still find the results of our derivation useful. Thus, in Fig. 9-2A

$$v_o = v_{in} \left(\frac{-jX_C}{R - jX_C} \right)$$

$$K = \frac{v_o}{v_{in}} = \frac{-jX_C}{R - jX_C} = \frac{X_C}{\sqrt{R^2 + X_C^2}} \angle -90° + \arctan\frac{X_C}{R}$$

Dividing numerator and denominator by X_C yields

$$K = \frac{1}{\frac{1}{X_C}\sqrt{R^2 + X_C^2}} \angle -90° + \arctan\frac{X_C}{R}$$

$$= \frac{1}{\sqrt{(R^2/X_C^2) + 1}} \angle -90° + \arctan\frac{X_C}{R}$$

Substituting $1/\omega C$ for X_C yields

$$K = \frac{1}{\sqrt{(\omega RC)^2 + 1}} \angle -90° + \arctan\frac{1}{\omega RC} \tag{9-6}$$

where $\omega = 2\pi f$.

As you can see, in Equation 9-6 the gain (K) is a complex (not complicated) quantity. The impedance and phasor diagram for this low-pass filter are shown in Fig. 9-3. You should note the following:

1. The cutoff frequency is

$$f_c = \frac{1}{2\pi RC} \tag{9-5}$$

2. The magnitude of the gain at any frequency is

$$|K| = \frac{1}{\sqrt{(\omega RC)^2 + 1}} \tag{9-7}$$

3. The output voltage will *lag* the input voltage by an angle ϕ, where

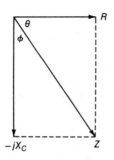

(A) Impedance diagram.

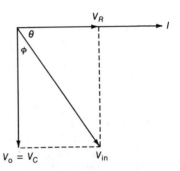

(B) Phasor diagram.

Fig. 9-3. Impedance and phasor diagrams for a low-pass RC filter.

$$\phi = -90° + \arctan\frac{1}{\omega RC} \qquad (9\text{-}8)$$

EXAMPLE 9-1

Use Equations 9-7 and 9-8 to determine the gain $|K|$ and the angle (θ) between the input and output voltage when $f = f_c$ for a low-pass RC filter.

$$f_c = \frac{1}{2\pi RC}$$

Therefore

$$\omega C = 2\pi f_c = \frac{1}{RC}$$

Substituting $\omega = 1/RC$ into Equation 9-7 yields

$$K = \frac{1}{\sqrt{[(1/RC) \times RC]^2 + 1}} = \frac{1}{\sqrt{2}} = 0.707$$

Substituting $\omega = 1/RC$ into Equation 9-8 yields

$$\phi = -90° + \arctan\frac{1}{(1/RC) \times RC}$$
$$= -90° + \arctan 1 = -45°$$

By employing Equations 9-7 and 9-8 we can predict the gain $|K|$ and angle ϕ between the input and output voltage for *any* frequency. It is very useful to construct a table of $|K|$, $|K_{dB}|$, and ϕ for values of f which are some multiple n of the cutoff frequency f_c. Thus at f_c we can write $\omega = \omega_c$ so that

$$\omega_c = 2\pi f_c = 2\pi\left(\frac{1}{2\pi RC}\right) = \frac{1}{RC}$$

When $f = nf_c$,

$$\omega = n\omega_c = \frac{n}{RC}$$

Substituting n/RC for ω in Equation 9-7 yields

$$|K| = \frac{1}{\sqrt{[(n/RC) \times RC]^2 + 1}} = \frac{1}{\sqrt{n^2 + 1}} \qquad (9\text{-}9)$$

What is ϕ when $f = nf_c$? Again, when $f = nf_c$

$$\omega = \frac{n}{RC}$$

Substituting n/RC for ω in Equation 9-8 yields

$$\phi = -90° + \arctan \frac{1}{\omega RC}$$

$$= -90° + \arctan \frac{1}{(n/RC) \times RC}$$

$$= -90° + \arctan \frac{1}{n} \qquad (9\text{-}10)$$

where in Equations 9-9 and 9-10 $n = f/f_c$.

Equations 9-9 and 9-10 enable us to construct Table 9-1, which provides values for $|K|$, $|K_{dB}|$, where $K_{dB} = 20 \log |K|$, and ϕ for frequencies well below, near, and well above the cutoff frequency f_c. Fig. 9-4 contains graphs of the data in Table 9-1. Since n represents "normalized" frequency the curves are applicable to *any* passive RC low-pass filter. With respect to Fig. 9-4 we note the following:

1. $|K_{dB}|$ is approximately zero for frequencies below the cutoff frequency.
2. Above the cutoff frequency the gain rolls off at approximately 6 dB/octave, which is equivalent to 20 dB/decade. (An octave is a 2:1 frequency ratio; a decade is a 10:1 frequency ratio.)
3. For frequencies well below cutoff the angle ϕ is approximately zero.
4. For frequencies well above cutoff the angle ϕ is approximately $-90°$.

These observations permit us to sketch the simplified graphs in Fig. 9-5. Graphs such as those in Fig. 9-5 are called *Bode diagrams* or *Bode plots*. For frequencies well below and well above cutoff Bode plots are quite accurate. In these plots the greatest error occurs *at* the cutoff frequency. Thus if you want very accurate values of $|K|$ or ϕ for frequencies near f_c you should use Table 9-1 or Equations 9-7 and 9-8.

EXAMPLE 9-2

Design an RC low-pass filter that has a cutoff frequency of 47.8 Hz. Estimate the decibel gain and the angle between input and

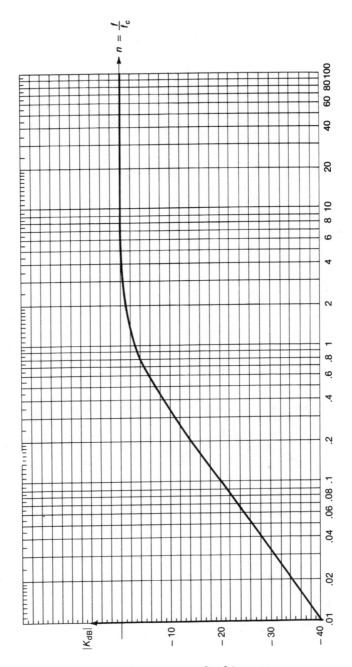

(A) Decibel gain vs. normalized frequency.

Fig. 9-8. High-pass RC

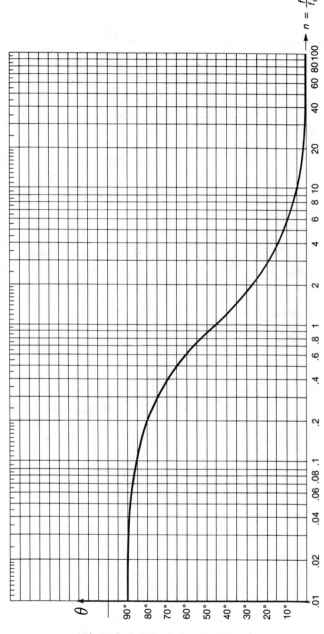

(B) Angle φ vs. normalized frequency.

filter characteristics.

Table 9-1. Low-Pass Filter Characteristics

f/f_c	$|K|$	$|K_{dB}|$ (dB)	$\phi(°)$
0.01	0.999	−0.0004	−0.573
0.1	0.995	−0.043	−5.711
0.2	0.981	−0.170	−11.310
0.3	0.958	−0.374	−16.700
0.4	0.929	−0.645	−21.801
0.5	0.894	−0.969	−26.565
0.6	0.858	−1.335	−30.964
0.7	0.819	−1.732	−34.992
0.8	0.781	−2.148	−38.660
0.9	0.743	−2.577	−41.987
1	0.707	−3.010	−45.000
2	0.447	−6.990	−63.435
4	0.243	−12.305	−75.964
6	0.164	−15.682	−80.534
8	0.124	−18.129	−82.875
10	0.099	−20.043	−84.289
100	0.009	−40.000	−89.430

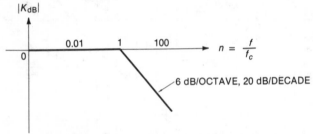

(A) Approximate decibel gain vs. normalized frequency.

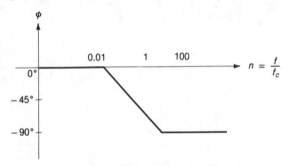

(B) Approximate angle ϕ vs. normalized frequency.

Fig. 9-5. Bode diagrams for a low-pass RC filter.

output voltages for the following frequencies: 4.78 Hz, 47.8 Hz, 95.6 Hz, and 4.78 kHz.

The cutoff frequency is

$$f_c = \frac{1}{2\pi RC} = 47.8 \text{ Hz}$$

Thus

$$RC = \frac{1}{2\pi f_c} = \frac{1}{2\pi(47.8)} = 3.33 \text{ ms}$$

There are of course an infinite number of RC combinations that yield the desired time constant. Somewhat arbitrarily, then, we select $C = 1\ \mu\text{F}$. Thus

$$R = \frac{3.33 \text{ ms}}{C} = \frac{3.33 \text{ ms}}{1\ \mu\text{F}} = 3.33 \text{ k}\Omega$$

In a practical design it is of course desirable to select standard values for R and C if at all possible.

For an approximate analysis of the filter response, refer to the Bode diagrams in Fig. 9-5. For this filter we would predict a gain of 0 dB up to cutoff (47.8 Hz). The gain would then roll off at 6 dB/octave, which is equivalent to 20 dB/decade. Thus, for the frequencies 4.78 Hz, 47.8 Hz, 95.6 Hz, and 4.78 kHz, the approximate gains would be 0 dB, 0 dB, −6 dB, and −40 dB, respectively. A more accurate analysis is obtained by referring to Table 9-1 which predicts the following values of gain and phase angle:

$\lvert K_{dB}\rvert = -0.043$ dB, $\phi = -5.711°$	(4.78 Hz)
$\lvert K_{dB}\rvert = -3.010$ dB, $\phi = -45.000°$	(47.8 Hz)
$\lvert K_{dB}\rvert = -6.990$ dB, $\phi = -63.435°$	(95.6 Hz)
$\lvert K_{dB}\rvert = -40.000$ dB, $\phi = -89.430°$	(4.78 kHz)

Note that the Bode diagram predictions are in close agreement with Table 9-1 *for* frequencies well below and well above cutoff.

9-5 THE RC HIGH-PASS FILTER

An RC high-pass filter and frequency response curve are illustrated in Fig. 9-6. Note that the cutoff frequency is again defined as the frequency where the gain (K) is 0.707. The impedance and phasor diagrams for the high-pass filter are provided in Fig. 9-7. Note that the output voltage *leads* the input voltage by an angle θ. An analysis similar to the one provided for the RC low-pass filter yields the following results:

$$K = \frac{1}{\sqrt{1 + 1/(\omega RC)^2}} \angle \arctan \frac{1}{\omega RC} \qquad (9\text{-}11)$$

Thus the magnitude of the gain $\lvert K\rvert$ at any frequency is

$$\lvert K\rvert = \frac{1}{\sqrt{1 + 1/(\omega RC)^2}} \qquad (9\text{-}12)$$

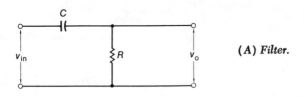

(A) Filter.

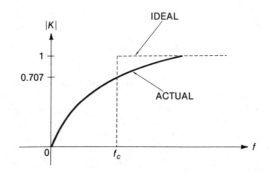

(B) Gain vs. frequency.

Fig. 9-6. A high-pass RC filter.

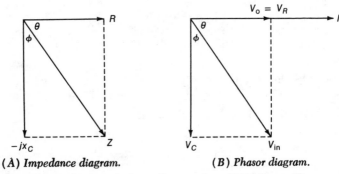

(A) Impedance diagram.   (B) Phasor diagram.

Fig. 9-7. Impedance and phasor diagrams for a RC high-pass filter.

Similarly, the angle (θ) by which the output voltage *leads* the input voltage is

$$\theta = \arctan \frac{1}{\omega RC} \qquad (9\text{-}13)$$

As was the case with the low-pass filter, the cutoff frequency and decibel gain are

$$f_c = \frac{1}{2\pi RC}$$

$$K_{dB} = 20 \log |K|$$

A table and graphs of $|K|$, $|K_{dB}|$ and θ can be constructed for values of f which are some multiple of f_c. Thus

$$|K| = \frac{1}{\sqrt{1 + 1/n^2}} \tag{9-14}$$

$$\theta = \arctan \frac{1}{n} \tag{9-15}$$

where $n = f/f_c$.

Equations 9-14 and 9-15 enable us to construct Table 9-2 and the curves provided in Fig. 9-8. Once again, for frequencies well below or well above cutoff you can accurately predict the response of the filter by employing the simplified graphs (Bode diagrams) provided in Fig. 9-9.

EXAMPLE 9-3

Design an RC high-pass filter that has a cutoff frequency of 1.59 kHz. Sketch a Bode diagram for $|K_{dB}|$, and estimate the output voltage when the input voltage is 10 V peak at 15.9 Hz.

Table 9-2. High-Pass Filter Characteristics

f/f_c	$|K|$	$|K_{dB}|$ (dB)	$\theta(°)$
0.01	0.009	−40.00	89.430
0.1	0.099	−20.043	84.289
0.2	0.196	−14.150	78.690
0.3	0.287	−10.830	73.301
0.4	0.371	−8.603	68.200
0.5	0.447	−6.990	63.435
0.6	0.515	−5.770	59.040
0.7	0.574	−4.830	55.010
0.8	0.625	−4.090	51.340
0.9	0.669	−3.490	48.010
1	0.707	−3.010	45.000
2	0.894	−0.969	26.565
4	0.970	−0.263	14.036
8	0.992	−0.067	7.125
10	0.995	−0.043	5.710
100	0.999	−0.0004	0.573

From the preceding discussion

$$RC = \frac{1}{2\pi f_c} = \frac{1}{2\pi (1.59 \text{ kHz})} = 0.1 \text{ ms}$$

Selecting $C = 0.01 \ \mu\text{F}$ we have

241

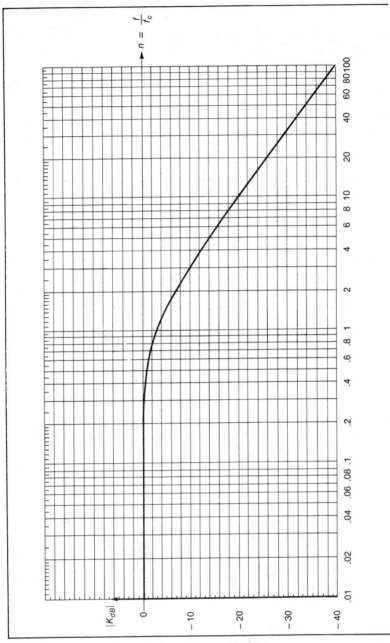

(A) Decibel gain vs. normalized frequency.

Fig. 9-4. Low-pass RC

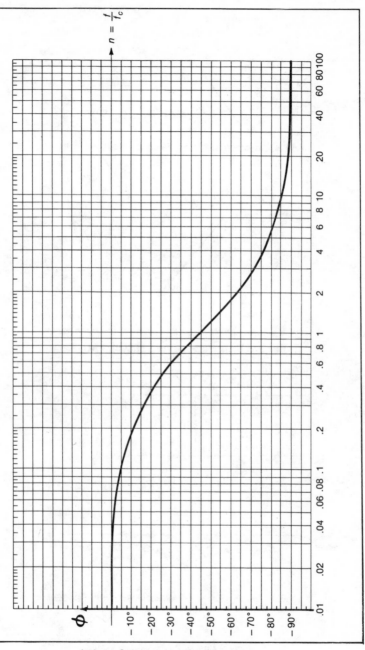

(B) Angle φ vs. normalized frequency.

filter characteristics.

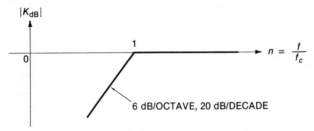

(A) Approximate decibel gain vs. normalized frequency.

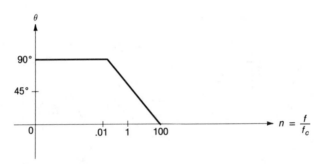

(B) Approximate angle θ vs. normalized frequency.

Fig. 9-9. Bode diagrams for a high-pass RC filter.

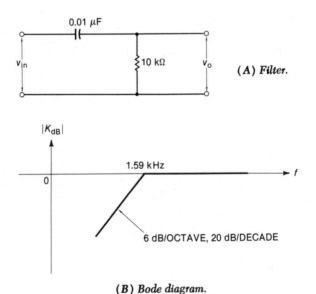

(A) Filter.

(B) Bode diagram.

Fig. 9-10. High-pass filter for Example 9-3.

$$R = \frac{0.1 \text{ ms}}{0.01 \text{ } \mu\text{F}} = 10 \text{ k}\Omega$$

The circuit and Bode diagram is illustrated in Fig. 9-10. When $f = 15.9$ Hz,

$$\frac{f}{f_c} = \frac{15.9 \text{ Hz}}{1.59 \text{ kHz}} = 0.01$$

For $f/f_c = 0.01$ Table 9-2 indicates that $|K| = 0.009$. Thus

$$\begin{aligned} v_o &= |K| \, v_{in} \\ &= 0.009(10 \text{ V}) \\ &= 0.09 \text{ V} = 90 \text{ mV peak} \end{aligned}$$

is the output voltage.

9-6 FILTER LOADING EFFECTS

The resistance of the signal source (R_s) driving a filter, and the resistance of the load (R_L) connected to the output of the filter, can significantly change the characteristics of the filter. *Specifically, the cutoff frequency and gain in the passband region are changed by the presence of source and load resistance.* Fig. 9-11A illustrates an RC low-pass filter where the effects of source and load resistance are

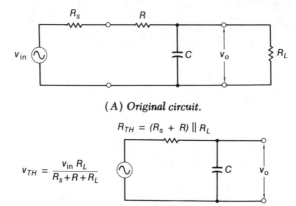

(A) *Original circuit.*

(B) *Thevenin equivalent.*

Fig. 9-11. Low-pass RC filter including source and load resistance.

to be considered. Fig. 9-11B is the Thevenin equivalent circuit for the filter in Fig. 9-11A. By utilizing the Thevenin equivalent circuit you can accurately predict the characteristics of the filter when the effects of source and load resistance are significant.

EXAMPLE 9-4

The low-pass filter in Example 9-2 had $R = 3.33$ kΩ, $C = 1$ μF, and $f_c = 47.8$ Hz. Estimate the cutoff frequency and passband gain if the filter is driven from a signal source whose resistance is 2 kΩ and terminated in a load resistance of 5 kΩ. See Fig. 9-12A.

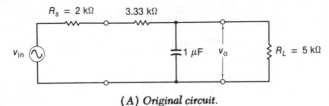

(A) Original circuit.

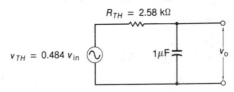

(B) Thevenin equivalent.

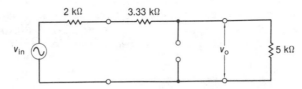

(C) Equivalent circuit in the passband region.

Fig. 9-12. Low-pass RC filter for Example 9-4.

We begin by determining the Thevenin voltage and resistance for the "loaded" filter:

$$v_{TH} = \frac{v_{in}R_L}{R_s + R + R_L} \tag{9-16}$$

$$= \frac{v_{in}(5\text{ kΩ})}{10.33\text{ kΩ}} = 0.484 v_{in}$$

$$R_{TH} = (R_s + R) \parallel R_L \tag{9-17}$$

$$= 5.33\text{ kΩ} \parallel 5\text{ kΩ} = 2.58\text{ kΩ}$$

See Fig. 9-12B.

Next we determine the cutoff frequency (f_c):

$$f_c = \frac{1}{2\pi R_{TH}C} \tag{9-18}$$

$$= \frac{1}{2\pi(2.58\text{ kΩ})(1\text{ μF})} = 61.63\text{ Hz}$$

For frequencies within the passband $|X_C|$ is very *large* compared with R_{TH}, and thus C approximates an *open circuit*. This enables us to visualize the equivalent circuit shown in Fig. 9-12C. Thus the gain in the passband region is approximated by

$$|K_{PB}| = \frac{R_L}{R_s + R + R_L} \tag{9-19}$$

$$= \frac{5\,k\Omega}{10.33\,k\Omega} = 0.484$$

$$|K_{PB(dB)}| = 20 \log |K_{PB}| = -6.3\,dB$$

The effects of source and load resistance are illustrated graphically by the Bode diagrams in Fig. 9-13. Note that the loaded filter does *not* have a passband gain of unity ($|K_{PB}| = 0.484 = -6.3\,dB$). Thus

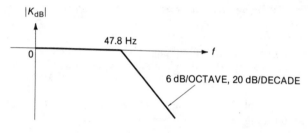

(A) *Unloaded filter.*

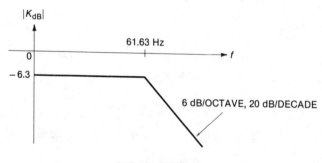

(B) *Loaded filter.*

Fig. 9-13. Bode diagrams for Example 9-4.

signals within the passband will be attenuated. This unwanted reduction in signal level is commonly referred to as *insertion loss* since it is a direct result of "inserting" a filter into a circuit.

Fig. 9-14 illustrates an RC high-pass filter with source and load resistance present. In order to predict the characteristics of the loaded filter we will again employ the Thevenin equivalent circuit.

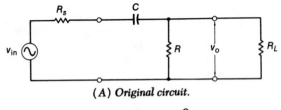

(A) Original circuit.

$$v_{TH} = v_{in} \left(\frac{R \parallel R_L}{R_s + R \parallel R_L} \right) \qquad R_{TH} = R_s + R \parallel R_L$$

(B) Thevenin equivalent.

Fig. 9-14. High-pass RC filter including source and load resistance.

EXAMPLE 9-5

The high-pass filter in Example 9-3 had $R = 10$ kΩ, $C = 0.01$ μF, and $f_c = 1.59$ kHz. Sketch Bode diagrams for the loaded and unloaded filter. For the loaded filter assume $R_s = 5$ kΩ and $R_L = 10$ kΩ.

Again we will first determine the Thevenin quantities for the loaded filter (Fig. 9-15):

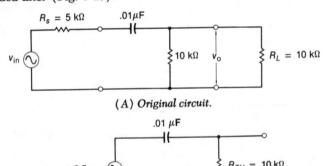

(A) Original circuit.

(B) Thevenin equivalent.

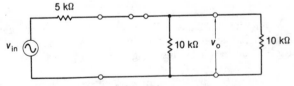

(C) Equivalent circuit in the passband region.

Fig. 9-15. High-pass RC filter for Example 9-5.

$$v_{TH} = v_{\text{in}}\left(\frac{R \parallel R_L}{R_s + R \parallel R_L}\right) \qquad (9\text{-}20)$$

$$R \parallel R_L = 10 \text{ k}\Omega \parallel 10 \text{ k}\Omega = 5 \text{ k}\Omega$$

$$v_{TH} = v_{\text{in}}\left(\frac{5 \text{ k}\Omega}{10 \text{ k}\Omega}\right) = 0.5 \, v_{\text{in}}$$

$$R_{TH} = R_s + R \parallel R_L \qquad (9\text{-}21)$$
$$= 5 \text{ k}\Omega + 5 \text{ k}\Omega = 10 \text{ k}\Omega$$

The cutoff frequency (f_c) is

$$f_c = \frac{1}{2\pi R_{TH} C} = \frac{1}{2\pi (10 \text{ k}\Omega)(0.01 \text{ }\mu\text{F})} = 1.59 \text{ kHz}$$

For frequencies within the passband $|X_C|$ is very *small* compared to R_{TH}. Thus for the high-pass filter C approximates a *short* circuit in the passband region. This is illustrated in Fig. 9-15C. Thus

$$|K_{PB}| = \frac{R \parallel R_L}{R_s + R \parallel R_L} \qquad (9\text{-}22)$$

$$= \frac{5 \text{ k}\Omega}{10 \text{ k}\Omega} = 0.5$$

$$|K_{PB(\text{dB})}| = 20 \log 0.5 = -6.02 \text{ dB}$$

Note that the cutoff frequencies for the loaded and unloaded filter in this example are equal. This unusual case is due to the fact that $R_{TH} = R$. Even though the presence of source and load resistance *did not* change the cutoff frequency, it *did* introduce a rather significant insertion loss of -6.02 dB. Fig. 9-16 illustrates the Bode diagrams for the loaded and unloaded filter.

9-7 NOTCH FILTERS

Fig. 9-17 represents a generalized ac bridge. Note that the bridge is driven by an ac source and that the elements forming the bridge arms are *impedances* rather than resistances. The analysis of ac bridges is similar to the analysis of dc bridges provided in Chapter 8. Since we are dealing with impedance rather than resistance, the algebra necessary to analyze an ac bridge is often quite "tedious." Ac bridges are employed to measure capacitance, inductance, and impedance in a manner similar to the way a Wheatstone bridge is used to measure resistance. Thus an ac bridge will be balanced when $Z_1 Z_4 = Z_2 Z_3$.

A Wien bridge, which is a type of ac bridge, is shown in Fig. 9-18. One of the interesting characteristics of a Wien bridge is that balance occurs for only *one* frequency. Specifically, the bridge will be balanced when $Z_1 Z_4 = Z_2 Z_3$. Thus, with reference to Fig. 9-18,

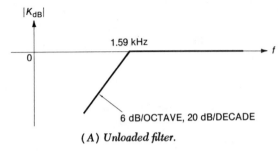

(A) *Unloaded filter.*

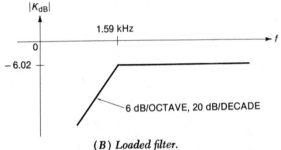

(B) *Loaded filter.*

Fig. 9-16. Bode diagrams for Example 9-5.

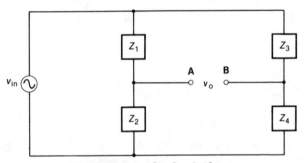

Fig. 9-17. Generalized ac bridge.

$$Z_1 = 2R_1$$

$$Z_2 = R_1$$

$$Z_3 = R - jX_C = R - \frac{j}{\omega C} = \frac{\omega RC - j}{\omega C}$$

$$Z_4 = \frac{R(-jX_C)}{R - jX_C} = \frac{-jR/\omega C}{(\omega RC - j)/\omega C}$$

$$= \frac{-jR}{\omega RC - j}$$

For balance, $Z_1Z_4 = Z_2Z_3$:

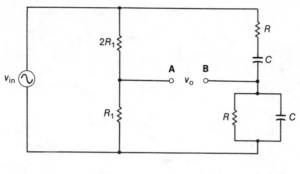

$$Z_1 = 2R_1 \qquad\qquad\qquad Z_3 = R - jX_C$$

$$Z_2 = R_1 \qquad\qquad\qquad Z_4 = R \,\|\, (-jX_C)$$

Fig. 9-18. A Wien bridge.

$$2R_1 \left(\frac{-jR}{\omega RC - j} \right) = R_1 \left(\frac{\omega RC - j}{\omega C} \right)$$

$$-j2R\omega C = (\omega RC - j)^2$$

$$-j2\omega RC = (\omega RC)^2 - j2\omega RC - 1$$

$$(\omega RC)^2 = 1$$

$$\omega RC = 1$$

$$2\pi f_c RC = 1$$

$$f_c = \frac{1}{2\pi RC} \qquad\qquad (9\text{-}23)$$

This result (Equation 9-23) suggests one application for a Wien bridge—a *notch* filter, since the bridge will reject signals ($v_{AB} \cong 0$) at frequencies near the notch frequency ($1/2\pi RC$).

In order to acquire some insight into the operation of the Wien bridge in Fig. 9-18 consider the following:

1. Since the ratio of resistance on the *left* side of the bridge is 2:1, voltage v_A at *any* frequency will equal $\frac{1}{3}v_{in}$.
2. For *low* frequencies $|X_C| \gg R$ and thus approximates an *open* circuit.
3. For *high* frequencies $|X_C| \ll R$ and thus approximates a *short* circuit.

The preceding observations suggest the low- and high-frequency equivalent circuits illustrated in Fig. 9-19. Thus for both low and high frequencies $v_B \cong 0$. Since $v_o = v_A - v_B$ the output voltage will be approximately $\frac{1}{3}v_{in}$, *in the passband region.* This corresponds to an insertion loss of approximately −9.54 dB.

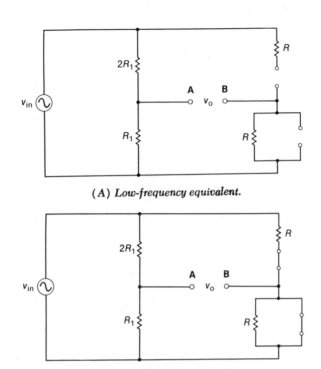

(A) Low-frequency equivalent.

(B) High-frequency equivalent.

Fig. 9-19. Approximate low- and high-frequency models for a Wien bridge.

Fig. 9-20 illustrates the gain vs. frequency response for the Wien-bridge circuit. In order to determine how the gain rolls off near the notch frequency it is (alas) necessary to derive an equation that predicts the gain at *any* frequency. Thus with reference to Fig. 9-18 we flex our algebraic muscles and proceed as follows:

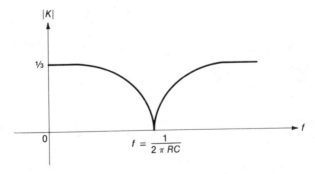

Fig. 9-20. Gain vs. frequency for the Wien bridge in Fig. 9-18.

$$v_B = v_{\text{in}}\left(\frac{Z_4}{Z_3 + Z_4}\right)$$

In complex notation,

$$v_B = v_{\text{in}}\left(\frac{\dfrac{-jR}{\omega RC - j}}{\dfrac{\omega RC - j}{\omega C} + \dfrac{-jR}{\omega RC - j}}\right)$$

$$= v_{\text{in}}\left[\frac{(\omega RC - j)\dfrac{-jR}{\omega RC - j}}{(\omega RC - j)\dfrac{(\omega RC - j)}{\omega C} + \dfrac{-jR(\omega RC - j)}{\omega RC - j}}\right]$$

$$= v_{\text{in}}\left[\frac{-jR}{\dfrac{(\omega RC)^2 - 2j(\omega RC) - 1}{\omega C} - jR}\right]$$

$$= v_{\text{in}}\left[\frac{-jR}{\dfrac{(\omega RC)^2 - 2j(\omega RC) - 1 - j(\omega RC)}{\omega C}}\right]$$

$$= v_{\text{in}}\left[\frac{-j(\omega RC)}{(\omega RC)^2 - 1 - j3(\omega RC)}\right]$$

Recall that $v_A = v_{\text{in}}/3$, and $v_o = v_A - v_B$. Therefore

$$K = \frac{v_o}{v_{\text{in}}} = \frac{1}{3} - \left[\frac{-j(\omega RC)}{(\omega RC)^2 - 1 - j3(\omega RC)}\right]$$

$$= \frac{(\omega RC)^2 - 1 - j3(\omega RC) + j3(\omega RC)}{3[(\omega RC)^2 - 1 - j3(\omega RC)]}$$

$$= \frac{1}{3}\left[\frac{(\omega RC)^2 - 1}{(\omega RC)^2 - 1 - j3(\omega RC)}\right]$$

$$= \frac{1}{3}\left[\frac{\omega RC - \dfrac{1}{\omega RC}}{\omega RC - \dfrac{1}{\omega RC} - j3}\right]$$

Converting to polar notation we have the much sought after result:

$$K = \frac{\omega RC - \dfrac{1}{\omega RC}}{3\sqrt{\left(\omega RC - \dfrac{1}{\omega RC}\right)^2 + 9}} \angle \arctan\frac{3}{\omega RC - 1/\omega RC} \quad (9\text{-}24)$$

Fig. 9-21 illustrates decibel gain vs. normalized frequency for the Wien-bridge circuit. Note that for frequencies less than one tenth

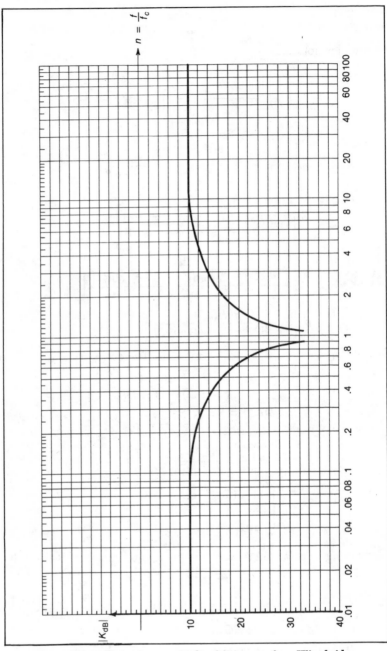

Fig. 9-21. Decibel gain vs. normalized frequency for a Wien bridge
(notch filter).

254

the notch frequency (low frequencies) the gain is relatively constant and is approximately −9.54 dB. Similarly, for frequencies greater than ten times the notch frequency (high frequencies) the gain is also approximately −9.54 dB. As you approach the notch frequency from either direction the gain rapidly decreases. Theoretically, the attenuation at the notch frequency approaches infinity. From a practical viewpoint, gains of −30 dB or more are quite common.

EXAMPLE 9-6

Design a Wien-bridge notch filter to reject the 60-Hz "hum" signal present at the input of a sensitive electronic system.

$$f_c = \frac{1}{2\pi RC}$$

$$RC = \frac{1}{2\pi f_c} = \frac{1}{2\pi(60)} = 2.65 \text{ ms}$$

Select $C = 1$ μF. Then

$$R = \frac{2.65 \text{ ms}}{1 \ \mu\text{F}} = 2.65 \text{ k}\Omega$$

A standard value 2.7-kΩ resistor would be acceptable for the design.

The resistance ratio for the left side of the bridge in Fig. 9-18 must be 2:1. Select $R_1 = 10$ kΩ; therefore $2R_1 = 20$ kΩ.

9-8 A BANDPASS FILTER

Fig. 9-22 illustrates a simple bandpass filter. As you can see, we have cascaded a low-pass and high-pass filter to obtain the bandpass filter. The circuit in Fig. 9-22B will produce a satisfactory result *if* the upper cutoff frequency (f_2) is large compared with the lower cutoff frequency (f_1). In addition R_2 should be large compared to R_1 to obtain the desired bandpass characteristic illustrated in Fig. 9-22C. Subject to the restrictions just described, the cutoff frequencies are (approximately)

$$f_1 = \frac{1}{2\pi R_2 C_2} \qquad (9\text{-}25)$$

$$f_2 = \frac{1}{2\pi R_1 C_1} \qquad (9\text{-}26)$$

EXAMPLE 9-7

Design a bandpass filter so that $f_1 = 3.18$ kHz and $f_2 = 318$ kHz. Estimate the gain in the passband region.

$$f_1 = \frac{1}{2\pi R_2 C_2} = 3.18 \text{ kHz}$$

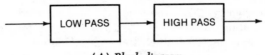

(A) Block diagram.

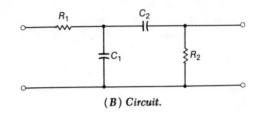

(B) Circuit.

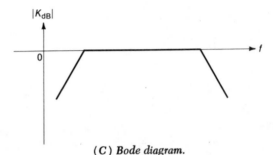

(C) Bode diagram.

Fig. 9-22. A bandpass filter.

$$R_2C_2 = \frac{1}{2\pi(3.18 \text{ kHz})} = 0.05 \text{ ms}$$

Select $R_2 = 0.5 \text{ M}\Omega$. Then

$$C_2 = \frac{0.05 \text{ ms}}{0.5 \text{ m}\Omega} = 0.1 \times 10^{-9} \text{ F} = 100 \text{ pF}$$

Now

$$f_2 = \frac{1}{2\pi R_1 C_1} = 318 \text{ kHz}$$

$$R_1C_1 = \frac{1}{2\pi(318 \text{ kHz})} = 0.5 \text{ } \mu s$$

Select $R_1 = R_2/10 = 50 \text{ k}\Omega$. Then

$$C_1 = \frac{0.5 \text{ } \mu s}{50 \text{ k}\Omega} = 0.01 \times 10^{-9} \text{ F} = 10 \text{ pF}$$

In the passband region C_1 appears open and C_2 shorted. Thus the passband gain is approximately

$$|K_{PB}| = \frac{R_2}{R_1 + R_2} = \frac{0.5 \text{ M}\Omega}{0.5 \text{ M}\Omega + 50 \text{ k}\Omega} = 0.909$$

which corresponds to −0.828 dB.

9-9 REVIEW OF OBJECTIVES

Filters pass certain frequencies and reject other frequencies. Unloaded high-pass and low-pass filters have passband gains near unity (0 dB), and a cutoff frequency of $1/2\pi RC$. The presence of source and load resistance can significantly change the passband gain and cutoff frequency. By employing the Thevenin equivalent circuit for a filter you can accurately predict the effects of source and load resistance.

The Wien bridge is often used as a *notch* filter since it is balanced at only one frequency. By cascading a low-pass and a high-pass filter you obtain a bandpass filter.

9-10 QUESTIONS

1. Discuss the characteristics of the four basic filter types.
2. Where is the greatest error on a Bode diagram? How much is this error?
3. What factors determine the passband gain and cutoff frequency for an *RC* low-pass filter?
4. How is the cutoff frequency of an ideal filter defined? A real filter?
5. What is meant by an octave? A decade?
6. Above the cutoff frequency, how rapidly does the gain of an *RC* low-pass filter decrease?
7. Dr. Courtine suggests using the bandpass filter in Fig. 9-22 to process signals whose frequencies are between 10 kHz and 20 kHz. Will the good doctor's suggestion work? Why?

9-11 PROBLEMS

1. Determine the passband gain and cutoff frequency for the filter in Fig. 9-23A.
2. Determine the decibel gain for the filter in Fig. 9-23A if the input signal is a 10-V peak sinusoid whose frequency is 95.4 Hz.
3. What is the angle between the input and output voltage in Problem 2?
4. Estimate the passband gain for the filter in Fig. 9-23B.
5. What will be the voltage across the 5-kΩ resistor in Fig. 9-23B when the input voltage is a 20-V peak sinusoid whose frequency is 100 kHz?
6. Design a high-pass *RC* filter that has the following specifications:
   Source resistance, 600 Ω
   Load resistance, 1 kΩ
   Cutoff frequency, 1 kHz
7. Repeat Problem 6 for a low-pass *RC* filter.
8. Estimate the decibel gain of a Wien-bridge notch filter when $f = 3f_c$.
9. Design the filter in Fig. 9-23C so that $f_1 = 15.9$ kHz and $f_2 = 159$ kHz. What decibel gain would you expect in the passband region?

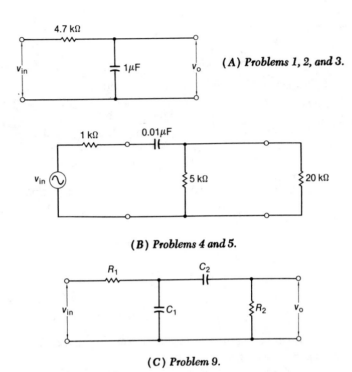

(A) Problems 1, 2, and 3.

(B) Problems 4 and 5.

(C) Problem 9.

Fig. 9-23. Schematic diagrams for Chapter 9 problems.

9-12 EXPERIMENT 9-1

Objective

The objective of this experiment is to design and evaluate the characteristics of an RC low-pass filter. Both loaded and unloaded filters will be considered.

Material Required

Resistor decades variable in 100-Ω steps (three required)
Capacitor (you will calculate the value required as part of the experiment)
Dual-trace oscilloscope
Function generator providing sinusoidal output voltages from 100 Hz to 100 kHz. The generator output impedance should equal 600 Ω

Introduction

A low-pass RC filter has a cutoff frequency specified by Equation 9-5. Specifically,

$$f_c = \frac{1}{2\pi RC} \qquad (9\text{-}5)$$

At the cutoff frequency the gain drops to 0.707 of the *passband* value, and the output voltage *lags* the input voltage by 45°. The Bode diagram of an *unloaded* filter indicates that the decibel gain is approximately 0 dB up to the cutoff frequency. Above the cutoff frequency the gain rolls off at 6 dB/octave, which is equivalent to 20 dB/decade.

The presence of source and/or load resistance can significantly alter the cutoff frequency and passband gain. When source resistance, load resistance, or both source and load resistance are present you should employ the Thevenin equivalent circuit of the filter to predict the cutoff frequency and passband gain. Specifically,

$$R_{TH} = (R_s + R) \parallel R_L \qquad (9\text{-}17)$$

$$f_c = \frac{1}{2\pi R_{TH}C} \qquad (9\text{-}18)$$

$$|K_{PB}| = \frac{R_L}{R_s + R + R_L} \qquad (9\text{-}19)$$

By utilizing the above relationships you can determine the characteristics of loaded *RC* low-pass filters.

Procedure

Step 1. Calculate the value of *C* required in Fig. 9-24A for a cutoff frequency of 3.18 kHz.

$$C = \underline{\hspace{2in}}$$

Step 2. Build the circuit in Fig. 9-24A. Connect the function generator to the input, and *then* adjust the output of the function generator for 1 V peak.

Step 3. Use an oscilloscope to measure the filter input voltage (channel 1), and output voltage (channel 2). Vary the frequency until $v_o = 0.707$ V. Record the *measured* cutoff frequency and phase angle.

$$f_c = \underline{\hspace{1.5in}} \qquad \phi = \underline{\hspace{1.5in}}$$

Step 4. Table 9-3 indicates calculated values of $|K|$ and $|K_{dB}|$. Recall that for an unloaded *RC* low-pass filter

$$|K| = \frac{1}{\sqrt{n^2 + 1}} \qquad (9\text{-}9)$$

and

$$|K_{dB}| = 20 \log |K|$$

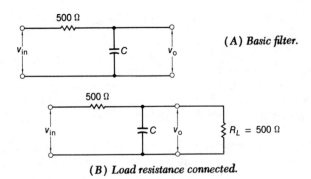

(A) Basic filter.

(B) Load resistance connected.

(C) Source resistance included.

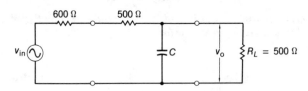

(D) Source and load resistance included.

Fig. 9-24. Schematic diagrams for Experiment 9-1.

Table 9-3. Data for the Unloaded Filter in Fig. 9-24A

f/f_o	f	$|K|_{CAL}$	$|K|_{MEAS}$	$|K_{dB}|_{CAL}$	$|K_{dB}|_{MEAS}$
0.01		0.995		−0.043	
1		0.707		−3.010	
2		0.447		−6.990	
4		0.243		−12.305	
6		0.164		−15.682	
8		0.124		−18.129	
10		0.099		−20.043	

where $n = f/f_c$. Equation 9-9 was used to generate the calculated values of gain in Table 9-3. Using the value of f_c measured in Step 3, complete the f column in Table 9-3 ($f = nf_c$).

Step 5. For each value of f in Table 9-3 measure v_o. Since the input voltage is 1 V peak the measured values of v_o equal the gain ($|K|_{MEAS}$). Each time you change frequency *make sure* $v_{in} = 1$ V peak. This will eliminate the effects of the generator source resistance. Complete the last column in Table 9-3: $|K_{dB}|_{MEAS} = 20 \log |K|_{MEAS}$.

Step 6. Repeat Steps 2 through 5 for the circuit shown in Fig. 9-24B. When you calculate $|K|_{CAL}$ remember to *include* the effects of load resistance. Record your data and calculations in Table 9-4.

Table 9-4. Data When $R_L = 500 \, \Omega$ (Fig. 9-24B)

f/f_c	f	$|K|_{CAL}$	$|K|_{MEAS}$	$|K_{dB}|_{CAL}$	$|K_{dB}|_{MEAS}$
0.1					
2					
4					
6					
8					
10					

Table 9-5. Data When $R_s = 600 \, \Omega$

f/f_c	f	$|K|_{CAL}$	$|K|_{MEAS}$	$|K_{dB}|_{CAL}$	$|K_{dB}|_{MEAS}$
0.1					
1					
2					
4					
6					
8					
10					

Step 7. To *include* the effects of the generator's source resistance we will *modify Step 2* as follows. With the filter *disconnected* from the generator adjust the generator's output to 1 V peak. For the remainder of the experiment *do not readjust* the generator's output voltage.

Step 8. Build the circuit shown in Fig. 9-24C. When you calculate $|K|_{CAL}$ remember to include the effects of source resistance. Record your data and calculations in Table 9-5.

Step 9. Build the circuit shown in Fig. 9-24D. When you calculate $|K|_{CAL}$ remember to include the effects of source *and* load resistance. Record your data and calculations in Table 9-6.

Table 9-6. Data When $R_L = R_s = 600 \, \Omega$

f/f_o	f	$|K|_{CAL}$	$|K|_{MEAS}$	$|K_{dB}|_{CAL}$	$|K_{dB}|_{MEAS}$
0.1					
1					
2					
4					
6					
8					
10					

Conclusion

When are source and load resistance significant? Sketch Bode diagrams for each circuit employed in this experiment. Design a filter to provide a cutoff frequency of 3.18 kHz *when $R_s = R_L = 600 \, \Omega$.* Will the Bode diagram for this filter be similar to any of the Bode diagrams encountered in this experiment? Why?

CHAPTER 10

Attenuators

10-1 INTRODUCTION

Attenuators are used to reduce signal levels by a fixed (constant) amount. Attenuators also serve as impedance matching devices between the signal source and load. Ideally, an attenuator should provide the same amount of attenuation for all signal frequencies.

Attenuators are often used to transfer signal energy from a source to the input of an instrument. An oscilloscope probe is a good example of an attenuator's application. In this chapter we will examine the types of attenuators commonly found in instruments.

10-2 OBJECTIVES

At the end of this chapter you will be able to do the following:

- Design a passive, frequency-compensated, oscilloscope probe.
- Analyze and design symmetrical, π-type attenuators.
- Analyze and design symmetrical, T-type attenuators.
- Design a π- or T-type pad for a signal source.
- Leap tall buildings in a single bound.

10-3 THE L-TYPE ATTENUATOR

An L-type attenuator is illustrated in Fig. 10-1. The L-type attenuator is nothing more than a simple voltage divider. Thus

$$v_o = v_{\text{in}} \left(\frac{R_2}{R_1 + R_2} \right)$$

263

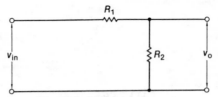

Fig. 10-1. An L-type attenuator.

Recall that attenuation (a) is defined as $a = v_{\text{in}}/v_o$. Therefore

$$a = \frac{R_1 + R_2}{R_2} \qquad (10\text{-}1)$$

and

$$a_{\text{dB}} = 20 \log a \qquad (10\text{-}2)$$

The design of an L-type attenuator is straightforward, as illustrated by Example 10-1.

EXAMPLE 10-1

Assume that R_2 in Fig. 10-1 is 1 MΩ. Design an attenuator to provide an attenuation of 20 dB. What is the output voltage if the input voltage is 1 V peak?

$$a_{\text{dB}} = 20 \log a$$
$$\log a = \frac{a_{\text{dB}}}{20} = \frac{20}{20} = 1$$

Thus

$$a = 10^1 = 10$$

Hence

$$a = \frac{R_1 + R_2}{R_2}$$
$$10 = \frac{R_1 + 1\,\text{MΩ}}{1\,\text{MΩ}}$$
$$R_1 + 1\,\text{MΩ} = 10\,\text{MΩ}$$
$$R_1 = 9\,\text{MΩ}$$
$$v_o = \frac{v_{\text{in}}}{a} = \frac{1}{10} = 0.1\,\text{V peak}$$

10-4 OSCILLOSCOPE PROBES

With the exception of ac bridges and filters, the measurement instruments discussed previously were assumed to have input impedances that were *purely resistive*. In the case of an oscilloscope this assumption is only valid for dc signals and low-frequency ac signals.

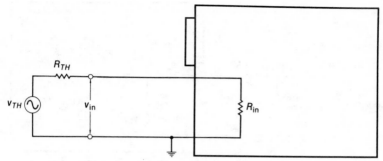

Fig. 10-2. Low-frequency (×1) probe and scope model.

Fig. 10-2 illustrates how a low-frequency ac signal is measured by an oscilloscope and a ×1 probe. The ×1 refers to the attenuation of the probe, which is unity. The function of the probe is to couple the signal to be measured to the input of the oscilloscope. Assuming $R_{in} >> R_{TH}$, very little loading occurs when the oscilloscope is placed in the circuit. Thus the waveform displayed on the oscilloscope will accurately represent the shape and amplitude of the signal being measured.

The "attenuator" in Fig. 10-2 is nothing more than a "pair of wires." Normally, these wires would be shielded (coaxial cable) to minimize picking up stray signals, noise, etc., from other sources. You can visualize the attenuator in Fig. 10-2 as an L-type attenuator in which $R_1 = 0$ and R_2 represents the input resistance of the scope (R_{in}).

Fig. 10-3 illustrates how a low-frequency ac signal is measured with a ×10 probe. Note that the ×10 probe is the L-type attenuator designed in Example 10-1. If the oscilloscope in Fig. 10-3 has a purely resistive input impedance of 1 MΩ, then the attenuation from the signal source to the scope's input would be 20 dB ($a = 10$) for *any* signal frequency. Unfortunately, this is *not* the case in practice because a "real scope" has a significant input capacitance as well as

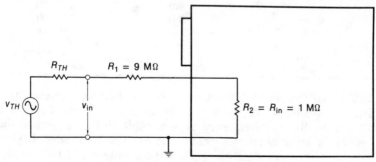

Fig. 10-3. Low-frequency (×10) probe and scope model.

265

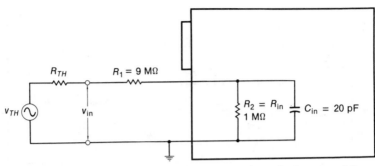

Fig. 10-4. Including the input capacitance in the probe/scope model.

input resistance. This is illustrated in Fig. 10-4. Note that the oscilloscope is assumed to have an input capacitance of 20 pF. How does the input capacitance affect the probe scope characteristics? Recall that the reactance of a capacitor varies with frequency. Specifically,

$$|X_{C(\text{in})}| = \frac{1}{2\pi f C_{\text{in}}} \tag{10-3}$$

At any frequency the input impedance of the scope consists of the parallel combination of R_{in} and $|X_{C(\text{in})}|$. Thus

$$Z_{\text{in}} = R_{\text{in}} \parallel X_{C(\text{in})}$$

$$|Z_{\text{in}}| = \frac{R_{\text{in}} X_{C(\text{in})}}{\sqrt{R_{\text{in}}^2 + X_{C(\text{in})}^2}} \tag{10-4}$$

By using Equations 10-3 and 10-4 for various frequencies, the data in Table 10-1 were obtained.

Table 10-1. Probe/Scope Input Impedance for Various Signal Frequencies

f	$|X_{C(\text{in})}|$	Z_{in}
100 Hz	80 MΩ	1 MΩ
1 kHz	8 MΩ	0.99 MΩ
10 kHz	800 kΩ	0.63 kΩ
100 kHz	80 kΩ	79.7 kΩ
1 MHz	8 kΩ	8 kΩ
10 MHz	800 Ω	800 Ω

Note that the input impedance of the probe/scope combination varies drastically with frequency. This also causes the attenuation to vary drastically with frequency since $a = (R_1 + Z_{\text{in}})/Z_{\text{in}}$ when the effect of the input capacitance of the oscilloscope is considered. Thus the probe/scope combination in Fig. 10-4 is only useful for measuring dc and low-frequency ac voltages.

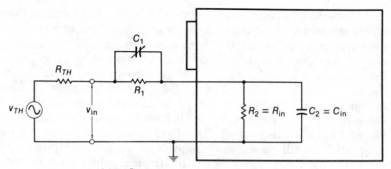

(A) The frequency-compensated probe.

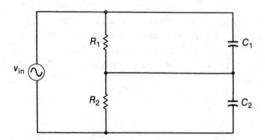

(B) Probe in (A) redrawn as an ac bridge.

Fig. 10-5. A frequency-compensated probe/scope combination.

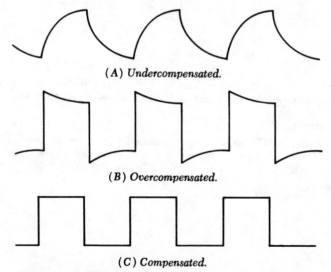

(A) Undercompensated.

(B) Overcompensated.

(C) Compensated.

Fig. 10-6. Adjusting the probe capacitance to obtain a properly compensated display.

If you wish to measure medium- and high-frequency signals, then a *frequency-compensated probe/scope combination* that provides a relatively constant attenuation over a wide range of frequencies is needed. An example of such a combination is illustrated in Fig. 10-5A. To understand how this probe/scope combination works, consider the equivalent circuit shown in Fig. 10-5B. Note that we can consider the probe/scope combination to be an ac bridge. By selecting a value for C_1 so that the bridge is "balanced," the attenuation is no longer frequency dependent. The input impedance of the scope $(X_{C(in)} \| R_{in})$ still varies with frequency, but *now* the probe impedance $(X_{C1} \| R_1)$ also varies with frequency—in such a manner that the value of $[(X_{C1} \| R_1) + (X_{C(in)} \| R_{in})]/(X_{C(in)} \| R_{in})$ remains constant. Since this value represents the attenuation of the frequency compensated probe/scope combination in Fig. 10-5A, the attenuation is no longer frequency dependent. How do you select a value for C_1? From our discussion of ac bridges, you should know that balance is achieved when $Z_1 Z_4 = Z_2 Z_3$. With reference to Fig. 10-5B we have

$$Z_1 = R_1, \quad Z_2 = R_2, \quad Z_3 = -j/\omega C_1, \quad Z_4 = -j/\omega C_2$$

Thus:

$$R_1 \left(\frac{-j}{\omega C_2} \right) = R_2 \left(\frac{-j}{\omega C_1} \right)$$
$$\frac{R_1}{C_2} = \frac{R_2}{C_1}$$

or

$$R_1 C_1 = R_2 C_2 \tag{10-5}$$

where $R_2 = R_{in}$ and $C_2 = C_{in}$. This result permits us to design frequency-compensated probes as illustrated by Example 10-2.

EXAMPLE 10-2

Design a ×5 frequency-compensated probe for an oscilloscope that has an input resistance of 1 MΩ shunted by 40 pF. First determine the value of R_1 required for an attenuation of 5 from Equation 10-1.

Equation 10-1 is

$$a = \frac{R_1 + R_2}{R_2}$$

so that

$$5 = \frac{R_1 + 1 \text{ M}\Omega}{1 \text{ M}\Omega}$$

$$R_1 + 1\ \text{M}\Omega = 5\ \text{M}\Omega$$
$$R_1 = 4\ \text{M}\Omega$$

Next, calculate the value of C_1 required to frequency compensate the probe/scope combination, from Equation 10-5:

$$R_1 C_1 = R_2 C_2$$
$$C_1 = \frac{R_2 C_2}{R_1} = \frac{(1\ \text{M}\Omega)\,(40\ \text{pF})}{4\ \text{M}\Omega} = 10\ \text{pF}$$

In a commercial probe, C_1 is made variable so that the probe can be used with more than one scope. The probe is calibrated to a particular scope by observing a square wave and adjusting C_1 until the best shape is obtained. The effect of adjusting C_1 can be seen in Fig. 10-6. Depending on the specific application an oscilloscope probe may itself become a rather sophisticated instrument. The point to keep in mind is simply this: the usefulness of an oscilloscope or any instrument is limited by the characteristics of the input coupling device.

10-5 CHARACTERISTIC IMPEDANCE

In the previous section of this chapter we saw a (compensation) network connected between the signal source and a load (contained in the oscilloscope). This is illustrated in a more general form in Fig. 10-7. The output impedance of the voltage source is R_{TH} and the load impedance is R_L. Depending on the specific application the

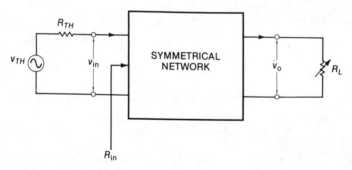

Fig. 10-7. General network connected between source and load.

network could be an amplifier, transmission line, filter, attenuator, or some other combination of passive and active elements. Since we are discussing attenuators we will assume the network in Fig. 10-7 is either a π- or T-type attenuator, as illustrated in Fig. 10-8. Note that the π-type attenuator is a delta (Δ) connection, and the T-type attenuator is a wye (Y) connection.

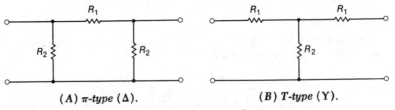

(A) π-type (Δ).   (B) T-type (Y).

Fig. 10-8. Two common types of attenuators.

The input impedance of an attenuator depends on the impedances used in the attenuator *and* the load impedance. Since the attenuators that we will discuss do not contain reactive elements and are terminated in a resistive load, their input impedance will be purely resistive (R_{in}).

Consider the circuit illustrated in Fig. 10-9. As R_L is varied, the input resistance of the attenuator changes. There will be *one* value of R_L that causes the input resistance of the attenuator to be *equal* to R_L and this is called the *characteristic resistance* of the attenuator (R_o). You can calculate the characteristic resistance of a symmetrical π- or T-type attenuator from

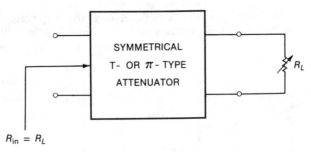

$R_{in} = R_L$

Fig. 10-9. The concept of characteristic resistance.

$$R_o = \sqrt{R_{in(o)} R_{in(s)}} \tag{10-6}$$

where

$R_{in(o)}$ = the input resistance of the attenuator when $R_L = \infty$,
$R_{in(s)}$ = the input resistance of the attenuator when $R_L = 0$.

EXAMPLE 10-3

Calculate the characteristic resistance of the π-type attenuator in Fig. 10-10A. Show that $R_{in} = R_o$ when $R_L = R_o$.

When $R_L = 0$ (Fig. 10-10B),

$$R_{in(s)} = 1\,800\ \Omega\ ||\ 448.8\ \Omega = 359.23\ \Omega$$

When $R_L = \infty$ (Fig. 10-10C),

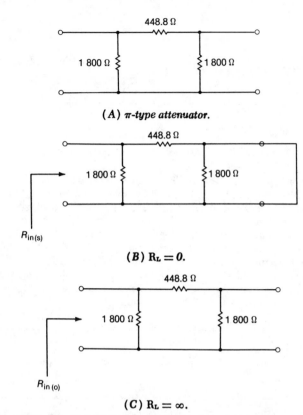

(A) π-type attenuator.

(B) $R_L = 0$.

(C) $R_L = \infty$.

Fig. 10-10. Circuits for calculating $R_{in(s)}$ and $R_{in(o)}$ for Example 10-3.

$$R_{in(o)} = 1\,800\,\Omega \parallel (448.8\,\Omega + 1\,800\,\Omega) = 999.8\,\Omega$$
$$R_o = \sqrt{R_{in(o)}R_{in(s)}}$$
$$= \sqrt{999.8\,(359.23)} = 600\,\Omega$$

Fig. 10-11 illustrates the attenuator of Fig. 10-10 terminated in a load resistance equal to R_o (600 Ω). Thus

$$R_{in} = 1\,800\,\Omega \parallel [448.8\,\Omega + (1\,800\,\Omega \parallel 600\,\Omega)]$$
$$= 1\,800\,\Omega \parallel (448.8\,\Omega + 450\,\Omega)$$
$$= 1\,800\,\Omega \parallel 898.8\,\Omega = 600\,\Omega$$

The point to remember from Example 10-3 is that *the input resistance of an attenuator equals R_o when the attenuator is terminated in a load resistance (R_L) equal to R_o.*

EXAMPLE 10-4

What is the input resistance of the circuit shown in Fig. 10-12?
Note that the circuit in Fig. 10-12 consists of two cascaded sections

271

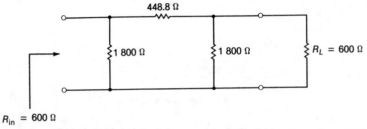

Fig. 10-11. The attenuator of Example 10-3 terminated with $R_L = R_o = 600 \, \Omega$.

of the π-type attenuator in Fig. 10-10A. Also note that the second section is terminated with a load resistance equal to the attenuator's characteristic resistance (600 Ω). Thus the input resistance of the second section will be 600 Ω. Since the input resistance of the second section constitutes the load resistance of the first section, the input resistance of the first section will also be 600 Ω. In fact no matter how many sections are cascaded the input resistance of the first section would still be 600 Ω! This assumes of course that the last section is terminated with a load resistance equal to the characteristic resistance of the attenuator (600 Ω).

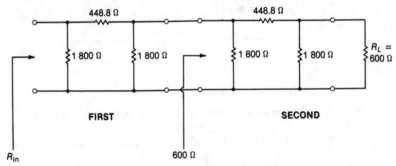

Fig. 10-12. Cascading two π sections.

Attenuators are normally used in *matched* systems as illustrated in Fig. 10-13. Note that in a matched system $R_{TH} = R_{in} = R_L = R_o$. A matched system has the following desirable characteristics:

1. Maximum transfer of *power* from the source to the attenuator.
2. Maximum transfer of *power* from the attenuator to the load.
3. The input resistance of *each section* in a series of cascaded attenuators is constant and equals R_o.

Recall that as the *frequency* (f) of an electromagnetic signal *increases* the *wavelength* (λ) *decreases*. Specifically,

$$\lambda = \frac{c}{f} \tag{10-7}$$

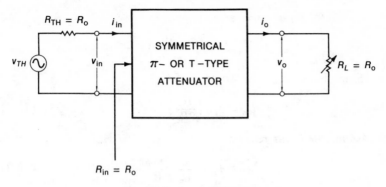

Fig. 10-13. A matched system.

where c is the velocity of light (approximately 3×10^8 m/s). Table 10-2 lists the wavelengths for the frequencies between 1 MHz and 1 THz. When the wavelength of a signal is of the same order of magnitude as the physical *length* between source and load, energy is lost in the form of radiation *unless* the system used for the transmission of the energy is matched. In effect, the conductors become antennas, radiating into space the energy you are trying to transfer from the source to the load. Thus it is especially important at high frequencies to use matched systems. The transmissions of computer data, radio signals, and television signals over telephones lines are all examples where matched systems are used. Thus the impedance matching function of attenuators is *very important*.

Table 10-2. Wavelengths for Frequencies Between 1 MHz and 1 THz

Frequency			Wavelength	
10^6	Hz =	1 MHz	300	m
10^7	Hz =	10 MHz	30	m
10^8	Hz =	100 MHz	3	m
10^9	Hz =	1 GHz	30	cm
10^{10}	Hz =	10 GHz	3	cm
10^{11}	Hz =	100 GHz	0.3	cm
10^{12}	Hz =	1 THz	0.03	cm

10-6 GENERAL ANALYSIS OF A MATCHED SYSTEM

In the next few sections you will learn how to analyze and design π- and T-type attenuators like those illustrated in Fig. 10-8. Before doing this, however, it will be instructive to first derive some basic relationships for the matched system of Fig. 10-13. Thus, with reference to Fig. 10-13 we proceed as follows.

1. **Attenuator input voltage:**

$$v_{in} = v_{TH}\left(\frac{R_{in}}{R_{TH} + R_{in}}\right)$$

Since $R_{TH} = R_{in} = R_o$ we have

$$v_{in} = v_{TH}\left(\frac{R_o}{2R_o}\right) = \frac{v_{TH}}{2} \tag{10-8}$$

2. **Attenuator input current:**

$$i_{in} = \frac{v_{in}}{R_{in}} = \frac{v_{in}}{R_o} \tag{10-9}$$

Since $v_{in} = v_{TH}/2$,

$$i_{in} = \frac{v_{TH}}{2R_o} \tag{10-10}$$

3. **Attenuator input power:**

$$P_{in} = v_{in}\,i_{in} = \left(\frac{v_{TH}}{2}\right)\left(\frac{v_{TH}}{2R_o}\right)$$

$$P_{in} = \frac{v_{TH}^2}{4R_o} \tag{10-11}$$

4. **Attenuator output voltage:**

$$a = \frac{v_{in}}{v_o}$$

Thus

$$v_o = \frac{v_{in}}{a} \tag{10-12}$$

$$v_o = \frac{v_{TH}}{2a} \tag{10-13}$$

5. **Attenuator output current:**

$$i_o = \frac{v_o}{R_L} = \frac{v_{in}}{aR_o} \tag{10-14}$$

$$i_o = \frac{v_{TH}}{2aR_o} \tag{10-15}$$

6. **Attenuator output (load) power:**

$$P_o = v_o i_o = \left(\frac{v_{in}}{a}\right)\left(\frac{v_{in}}{aR_o}\right)$$

$$P_o = \frac{v_{in}^2}{a^2 R_o} \tag{10-16}$$

$$P_o = \left(\frac{v_{TH}}{2a}\right)\left(\frac{v_{TH}}{2aR_o}\right) = \frac{v_{TH}^2}{4a^2R_o} \tag{10-17}$$

Note that Equations 10-8 through 10-17 are valid for virtually *any type* of attenuator. The only restriction is that the system in which the attenuator is placed be a matched system.

10-7   THE π-TYPE ATTENUATOR

A typical π-type attenuator (in a matched system) is shown in Fig. 10-14. Given values for R_1 and R_2, how would you determine the

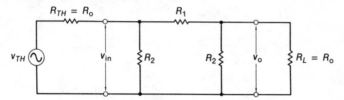

Fig. 10-14. A π-type attenuator in a matched system.

attenuation and characteristic resistance for the attenuator? We would use series parallel techniques to derive equations for R_o and a in terms of the component values. Such an analysis is not difficult but does involve a considerable amount of algebra. For this reason we will simply provide the desired results[*] For a π-type attenuator in a matched system

$$R_o = \frac{mR_1}{\sqrt{1 + 2m}} \tag{10-18}$$

$$a = \frac{1 + m + \sqrt{1 + 2m}}{m} \tag{10-19}$$

where $m = R_2/R_1$.

EXAMPLE 10-5

Determine the characteristic resistance and attenuation for the π-type attenuator in Fig. 10-15. Determine the input voltage, input current, and input power, assuming the attenuator is used in a matched system driven from a source whose Thevenin voltage is 10 V rms. Also determine the output voltage, output current, and output power.

---

[*] The analysis and design equations provided in this chapter are partially derived in Albert Malvino's *Electronic Instrumentation Fundamentals*, published by McGraw Hill.

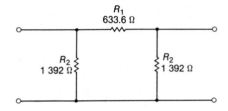

Fig. 10-15. A π-type attenuator for Example 10-5.

$$m = \frac{R_2}{R_1} = \frac{1\,392\ \Omega}{633.6\ \Omega} = 2.197$$

$$R_o = \frac{mR_1}{\sqrt{1+2m}}$$

$$R_o = \frac{2.197\,(633.6)}{\sqrt{1+2\,(2.197)}} = \frac{1\,392.02}{\sqrt{5.394}} = 600\ \Omega$$

Also,

$$a = \frac{1+m+\sqrt{1+2m}}{m}$$

$$= \frac{1+2.197+\sqrt{5.394}}{2.197} = 2.512$$

In terms of decibels,

$$a_{dB} = 20 \log a = 20 \log 2.512 = 8\ dB$$

Since $R_o = 600\ \Omega$, the attenuator should be used with a source whose output (Thevenin) resistance is 600 Ω and should be terminated in a load whose resistance is also 600 Ω. Assuming this is the case,

$$v_{in} = \frac{v_{TH}}{2} = \frac{10\ V}{2} = 5\ V\ rms$$

$$i_{in} = \frac{v_{in}}{R_o} = \frac{5\ V}{600\ \Omega} = 8.33\ mA\ rms$$

$$P_{in} = v_{in}i_{in} = (5\ V)\,(8.33\ mA) = 41.66\ mW$$

$$v_o = \frac{v_{in}}{a} = \frac{5\ V}{2.512} = 1.99\ V\ rms$$

$$i_o = \frac{v_o}{R_o} = \frac{1.99\ V}{600\ \Omega} = 3.32\ mA$$

$$P_o = v_oi_o = (1.99\ V)\,(3.32\ mA) = 6.61\ mW$$

Design is the opposite of analysis. When you design a π-type attenuator, your task is to determine values for R_1 and R_2 that provide the desired characteristic impedance and attenuation. The following equations are used to design π-type attenuators.

$$R_1 = \frac{R_0(a^2 - 1)}{2a} \tag{10-20}$$

$$R_2 = \frac{R_0(a + 1)}{a - 1} \tag{10-21}$$

EXAMPLE 10-6

Design a 600-Ω, 40-dB, π-type attenuator.

$$a_{dB} = 20 \log a = 40$$

$$\log a = \frac{40}{20} = 2$$

$$a = 10^2 = 100$$

$$R_1 = \frac{R_0(a^2 - 1)}{2a} = \frac{600(100^2 - 1)}{2(100)}$$

$$= 30\,000 \ \Omega = 30 \ \text{k}\Omega$$

$$R_2 = \frac{R_0(a + 1)}{a - 1} = \frac{600(100 + 1)}{100 - 1} = 612 \ \Omega$$

Fig. 10-16 illustrates the completed design. Naturally the attenuator should be used with a source and load resistance of 600 Ω.

Fig. 10-16. A 600-Ω, 40-dB, π-type
attenuator for Example 10-6.

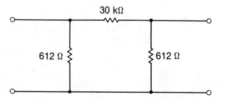

10-8 THE T-TYPE ATTENUATOR

The T-type attenuator can also be used in a matched system (Fig. 10-17). The equations that enable us to analyze and design a T-type attenuator are as follows:

$$R_0 = R_1 \sqrt{1 + 2m} \tag{10-22}$$

$$a = \frac{1 + m + \sqrt{1 + 2m}}{m} \tag{10-23}$$

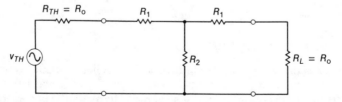

Fig. 10-17. A T-type attenuator in a matched system.

$$R_1 = \frac{R_o(a-1)}{a+1} \qquad (10\text{-}24)$$

$$R_2 = \frac{2aR_o}{a^2-1} \qquad (10\text{-}25)$$

EXAMPLE 10-7

Design a 10-dB, 50-Ω, T-type attenuator. Assume the attenuator is used in a properly matched system with a source whose Thevenin voltage is 5 V rms. Predict the output voltage and load power.

$$a_{dB} = 20 \log a = 10$$

$$\log a = \frac{10}{20} = 0.5$$

$$a = 10^{0.5} = 3.162$$

$$R_1 = \frac{R_o(a-1)}{a+1} = \frac{50(3.162-1)}{3.162+1} = 25.97 \ \Omega$$

$$R_2 = \frac{2aR_o}{a^2-1} = \frac{2(3.162)(50)}{(3.162)^2-1}$$

$$= \frac{316.2}{9} = 35.13 \ \Omega$$

$$v_o = \frac{v_{TH}}{2a} = \frac{5 \text{ V}}{2(3.162)} = 0.791 \text{ V rms}$$

$$P_o = \frac{v_{TH}^2}{4a^2R_o} = \frac{(5)^2}{4(3.162)^2(50)} = \frac{25}{1999.7}$$

$$= 12.5 \text{ mW}$$

The resulting design is straightforward (Fig. 10-18). In this case the attenuator should be used with a source and load resistance of 50 Ω. If an "exact" value of attenuation is not required, the closest standard resistor values would be employed in the design.

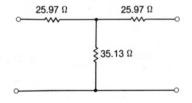

Fig. 10-18. A 50-Ω, 10-dB, T-type attenuator for Example 10-7.

10-9 PADDED SOURCES

Another name for an attenuator is a *pad*. When a signal source is connected to an attenuator, the source is said to be padded (Fig. 10-19). The purpose of padding a source is to make the output resistance of the padded source (R_{out}) equal to R_o. As you will soon see,

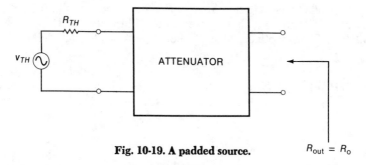

Fig. 10-19. A padded source.

$R_{\text{out}} = R_o$

if a source is well padded, then even large changes in the Thevenin resistance of the signal source (R_{TH}) will *not* significantly change the output resistance of the padded source (R_{out}). A π-type attenuator is shown in Fig. 10-20 padding a signal source. If R_{TH} varies over

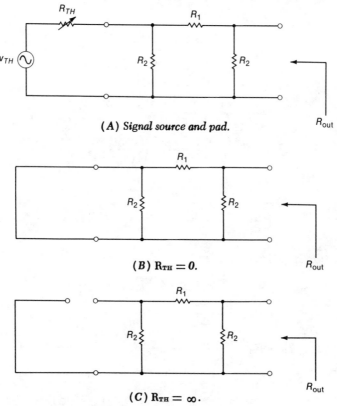

(A) Signal source and pad.

R_{out}

(B) $R_{TH} = 0$.

R_{out}

(C) $R_{TH} = \infty$.

R_{out}

Fig. 10-20. A π-type attenuator used to pad a source.

a wide range, what will the range in R_{out} be? We know that R_{TH} must be between 0 and ∞. These extremes are illustrated in Figs. 10-20B and 10-20C, respectively. Thus, when $R_{TH} = 0$ (Fig. 10-20B)

$$R_{out} = R_1 \parallel R_2$$

Recall that for a π-type attenuator,

$$R_1 = \frac{R_o(a^2 - 1)}{2a}$$

$$R_2 = \frac{R_o(a + 1)}{a - 1}$$

Therefore

$$R_{out} = \frac{\dfrac{R_o(a^2 - 1)}{2a} \times \dfrac{R_o(a + 1)}{a - 1}}{\dfrac{R_o(a^2 - 1)}{2a} + \dfrac{R_o(a + 1)}{a - 1}}$$

$$= \frac{\dfrac{R_o^2(a^2 - 1)(a + 1)}{2a(a - 1)}}{\dfrac{R_o(a^2 - 1)(a - 1) + 2aR_o(a + 1)}{2a(a - 1)}}$$

$$= \frac{R_o^2(a + 1)(a - 1)(a + 1)}{R_o[(a + 1)(a - 1)(a - 1) + 2a(a + 1)]}$$

$$= \frac{R_o(a + 1)(a - 1)(a + 1)}{(a + 1)[2a + (a - 1)(a - 1)]}$$

$$= \frac{R_o(a - 1)(a + 1)}{2a + a^2 - 2a + 1}$$

$$R_{out} = \frac{R_o(a^2 - 1)}{a^2 + 1} \tag{10-26}$$

Equation 10-26 is valid when $R_{TH} = 0$. When $R_{TH} = \infty$ (Fig. 10-20C)

$$R_{out} = (R_1 + R_2) \parallel R_2$$

$$R_1 + R_2 = \frac{R_o(a^2 - 1)}{2a} + \frac{R_o(a + 1)}{a - 1}$$

$$R_1 + R_2 = \frac{2aR_o(a + 1) + R_o(a^2 - 1)(a - 1)}{2a(a - 1)}$$

$$R_1 + R_2 = \frac{2aR_o(a + 1) + R_o(a + 1)(a - 1)^2}{2a(a - 1)}$$

Hence

$$R_{\text{out}} = \cfrac{\dfrac{2aR_o(a+1) + R_o(a+1)(a-1)^2}{2a(a-1)} \times \dfrac{R_o(a+1)}{a-1}}{\dfrac{2aR_o(a+1) + R_o(a+1)(a-1)^2}{2a(a-1)} + \dfrac{R_o(a+1)}{a-1}}$$

$$= \cfrac{\dfrac{2aR_o^2(a+1)^2 + R_o^2(a+1)^2(a-1)^2}{2a(a-1)^2}}{\dfrac{2aR_o(a+1) + R_o(a+1)(a-1)^2 + 2aR_o(a+1)}{2a(a-1)}}$$

$$= \frac{R_o^2(a+1)^2[2a + (a-1)^2]}{R_o(a+1)[2a + (a-1)^2 + 2a](a-1)}$$

$$= \frac{R_o(a+1)(2a + a^2 - 2a + 1)}{(a-1)(4a + a^2 - 2a + 1)}$$

$$= \frac{R_o(a+1)(a^2+1)}{(a-1)(a^2 + 2a + 1)}$$

$$= \frac{R_o(a+1)(a^2+1)}{(a-1)(a+1)^2}$$

Finally!

$$R_{\text{out}} = \frac{R_o(a^2+1)}{a^2 - 1} \qquad (10\text{-}27)$$

Equation 10-27 is valid *when $R_{TH} = \infty$*. Equations 10-26 and 10-27 enable us to predict the range in R_{out} for variations in R_{TH} between 0 and ∞. By using these equations, Table 10-3 was generated.

Table 10-3. The Output Resistance of a Padded Source

Attenuation (a)	R_{out} When $R_{TH} = 0$	R_{out} When $R_{TH} = \infty$
2	$0.600R_o$	$1.67R_o$
4	$0.882R_o$	$1.13R_o$
6	$0.946R_o$	$1.06R_o$
8	$0.969R_o$	$1.03R_o$
10	$0.980R_o$	$1.02R_o$
20	$0.995R_o$	$1.01R_o$

Look at Table 10-3 carefully. Note that when the attenuation of the pad is 10 or more ($a_{\text{dB}} = 20$) the output resistance of the padded source is between 98 and 102 percent of R_o, *even for* an *infinite* variation in R_{TH}! Suppose the value of R_{TH} is somewhere between 0 and ∞—how can you predict R_{out}? A good place to start would be with Fig. 10-20A, where you can see

$$R_{\text{out}} = [(R_{TH} \parallel R_2) + R_1] \parallel R_2$$

where R_{TH} is some multiple of R_o, i.e., $R_{TH} = nR_o$. Since the algebra required to derive a useful expression for R_{out} is "tedious" we will simply provide the result. Thus

$$R_{out} = \frac{R_o(a^2 + k)}{a^2 - k} \qquad (10\text{-}28)$$

where
$k = (n - 1)/(n + 1)$,
$n = R_{TH}/R_o$.

Equation 10-28 is *valid* for *any value of* R_{TH} between 0 and ∞. It should be mentioned that while we employed a π-type attenuator for our derivations, Equations 10-25, 10-26, and 10-27 apply to *any* attenuator!

EXAMPLE 10-8

A source has an output resistance (R_{TH}) of 600 Ω. Design a 20-dB, T-type pad so that the source can be used with a 50-Ω load. What is the output resistance of the padded source if $R_{TH} = 0$, ∞, and 600 Ω?

Note that 20 dB corresponds to an attenuation of 10. For a T-type attenuator

$$R_1 = \frac{R_o(a - 1)}{a + 1} = \frac{50(9)}{11} = 40.9 \ \Omega$$

$$R_2 = \frac{2aR_o}{a^2 - 1} = \frac{2(10)(50)}{99} = 10.1 \ \Omega$$

For $R_{TH} = 0$,

$$R_{out} = \frac{R_o(a^2 - 1)}{a^2 + 1} = \frac{50(99)}{101} = 49 \ \Omega$$

For $R_{TH} = \infty$,

$$R_{out} = \frac{R_o(a^2 + 1)}{a^2 - 1} = \frac{50(101)}{99} = 51 \ \Omega$$

Finally, for $R_{TH} = 600 \ \Omega$,

$$R_{out} = \frac{R_o(a^2 + k)}{a^2 - k}$$

where

$$n = \frac{R_{TH}}{R_o} = \frac{600}{50} = 12$$

$$k = \frac{n - 1}{n + 1} = \frac{11}{13} = 0.846$$

Thus

$$R_{out} = \frac{50(100.846)}{99.154} = 50.85 \ \Omega$$

10-10 BALANCED ATTENUATORS

The T- and π-type attenuators discussed previously are examples of *symmetrical unbalanced* attenuators. The term *symmetrical* refers to the fact that the input and output resistances of the attenuator are equal. As you have seen, symmetrical attenuators are normally used between a source and load whose resistances are equal. The term *unbalanced* means that one line of the attenuator is grounded while the other line is above ground potential.

Since symmetrical unbalanced attenuators are the types of attenuators you are most likely to encounter in practice, we have discussed them in detail. Occasionally the need arises for a balanced attenuator. In a balanced attenuator *both lines* of the attenuator are *above* ground potential (Fig. 10-21). Note that each input line of this attenuator is above ground potential due to the fact that the input to

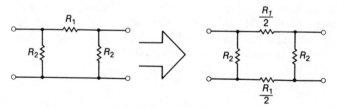

Fig. 10-21. One application for a balanced attenuator.

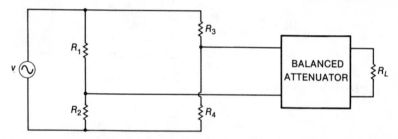

(A) Unbalanced T to balanced H.

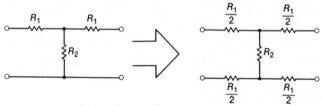

(B) Unbalanced π to balanced O.

Fig. 10-22. Converting symmetrical unbalanced T- and π-type attenuators to symmetrical balanced H- and O-type attenuators.

283

the attenuator comes from the output of a Wheatstone bridge. Additional examples where balanced attenuators would be used include two-wire balanced transmission lines, balanced generators, and balanced loads. Fig. 10-22 illustrates how symmetrical unbalanced T- and π-type attenuators can be converted to symmetrical balanced H- and O-type attenuators. The design of balanced attenuators is straightforward, as demonstrated by Example 10-9.

EXAMPLE 10-9

Design a balanced, 6-dB, 600-Ω, O-type attenuator. First determine the values of R_1 and R_2 for an unbalanced π-type attenuator.

$$a_{dB} = 20 \log a = 6$$

$$\log a = \frac{6}{20} = 0.3$$

Thus $a = 10^{0.3} = 1.995 \cong 2$. Also,

$$R_1 = \frac{R_o(a^2 - 1)}{2a} = \frac{600(3)}{4} = 450 \ \Omega$$

$$R_2 = \frac{R_o(a + 1)}{a - 1} = \frac{600(3)}{2} = 900 \ \Omega$$

For the balanced version

$$\frac{R_1}{2} = \frac{450 \ \Omega}{2} = 225 \ \Omega$$

This attenuator is shown in Fig. 10-23.

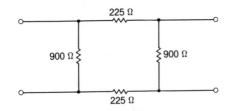

Fig. 10-23. A 600-Ω, 6-dB balanced O-type attenuator for Example 10-9.

Before concluding our discussion of attenuators we will briefly examine two additional types—the bridged-T (unbalanced) and bridged-H type (balanced). (Fig. 10-24). The analysis and design equation for a bridged-T type attenuator are:

$$R_o = R_1 \qquad (10\text{-}29)$$

$$a = \frac{R_1 + R_2}{R_2} \qquad (10\text{-}30)$$

$$R_2 = \frac{R_o}{a - 1} \qquad (10\text{-}31)$$

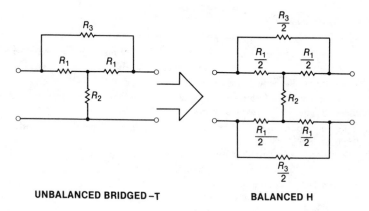

UNBALANCED BRIDGED–T　　　　　**BALANCED H**

Fig. 10-24. Converting a symmetrical unbalanced bridged-T type attenuator to a symmetrical balanced H-type attenuator.

$$R_3 = R_o(a - 1) \tag{10-32}$$

Since the bridged-T type attenuator has one more resistor than a T-type attenuator, perhaps you are wondering why it is used. The bridged-T type attenuator is often used in *variable* attenuators. Equation 10-30 indicates that you can vary the attenuation (a) by varying R_2. Since $R_3 = R_o(a - 1)$, when a is changed the value of R_3 must also be changed, for proper operation of the attenuator. Thus, R_2 and R_3 could be potentiometers that have their shafts ganged together when the bridged-T attenuator is employed as a variable attenuator.

10-11 REVIEW OF OBJECTIVES

Attenuators are used to reduce signal levels and match impedances. In a matched system $R_{TH} = R_o = R_L$. Matched systems are necessary to minimize power losses as the signal is transferred from the source to the load.

Since oscilloscopes have input capacitance as well as input resistance, it is necessary to frequency compensate the oscilloscope probe. A compensated probe ensures that attenuation from the signal source to the input of the oscilloscope will be constant over a wide frequency range.

Signal sources are often padded to stabilize the output resistance with respect to changes in the Thevenin resistance of the unpadded source. Table 10-3 illustrates how the effectiveness of a pad increases with attenuation. The design and analysis of common attenuator types were also illustrated, with an emphasis on symmetrical unbalanced T- and π-type attenuators.

10-12 QUESTIONS

1. What is meant by a *symmetrical* attenuator?
2. What is the difference between a balanced and unbalanced attenuator?
3. What does the term *characteristic resistance* mean?
4. Why are oscilloscope probes frequency compensated?
5. What procedure is followed to frequency compensate an oscilloscope probe?
6. What is meant by a *padded* source?
7. Does the input impedance of an oscilloscope increase or decrease with frequency? Why?
8. When does the input resistance of an attenuator equal R_o?
9. Under what conditions will a transmission line radiate energy?
10. When would a bridged-T type attenuator be preferred to a T-type attenuator? Would linear potentiometers be used? (Hint: look at Equations 10-23 and 10-30.)

10-13 PROBLEMS

1. Design a 72-Ω, 10-dB, π-type attenuator.
2. Convert the unbalanced attenuator in Problem 1 to a balanced attenuator.
3. Repeat Problems 1 and 2 for a T-type attenuator.
4. Calculate the values of R_1 and R_2 in Fig. 10-25A so that the attenuation is 8 dB. What is the power delivered to R_L if $R_{TH} = 50\ \Omega$?
5. Estimate the output resistance for the attenuator in Fig. 10-25A for the following values of R_{TH}: 0 Ω, 50 Ω, ∞. Assume $a = 8$.
6. The resistors in Fig. 10-25B have the following values:

    $R_1 = 129.06\ \Omega$
    $R_2 = 284.15\ \Omega$
    $R_3 = 179.52\ \Omega$
    $R_4 = 160.91\ \Omega$

    (*a*) Calculate the attenuation of each section.
    (*b*) Calculate the total attenuation.
    (*c*) Express the attenuations in (*a*) and (*b*) in decibels.
    (*d*) Calculate the values of v_1, v_2, and v_3.

7. Design a frequency-compensated $\times 5$ probe for an oscilloscope that has an input resistance of 1 MΩ shunted by 47 pF.
8. Design a balanced, H-type attenuator that meets the following specs:

    $R_o = 300\ \Omega$
    $a_{dB} = 12\ \Omega$

9. The circuit in Fig. 10-25C represents a voltage divider for the input to an instrument. The output voltage should be 100 V when the switch is in position 1, 50 V for position 2, and 5 V for position 3. Assuming $R_1 + R_2 + R_3 = 10$ MΩ, find R_1, R_2, and R_3.
10. Would it be possible to frequency compensate the voltage divider in Fig. 10-25C? What additional information would be required?

10-14 EXPERIMENT 10-1

Objective

The objective of this experiment is to investigate the characteristics of symmetrical, unbalanced, T-type attenuators.

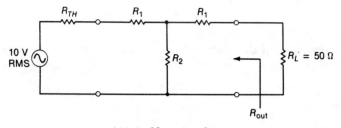

(A) Problems 4 and 5.

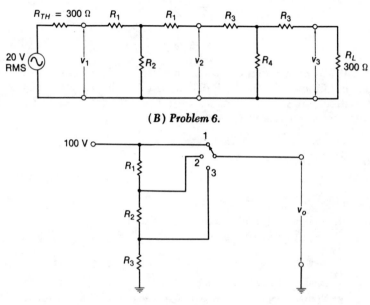

(B) Problem 6.

(C) Problem 9.

Fig. 10-25. Schematic diagrams for Chapter 10 problems.

Material Required

Resistor decades variable in 100-Ω, 10-Ω, 1-Ω, and 0.1-Ω steps (three required)

Signal generator with 600-Ω output impedance

Oscilloscope

Introduction

In a matched system ($R_{TH} = R_o = R_L$) you can calculate the input and output voltage of an attenuator by

$$v_{in} = \frac{v_{TH}}{2} \tag{10-8}$$

287

$$v_o = \frac{v_{in}}{a} = \frac{v_{TH}}{2a} \qquad (10\text{-}13)$$

The analysis and design equations for the T-type attenuator are as follows:

$$R_o = R_1\sqrt{1 + 2m} \qquad (10\text{-}22)$$

$$a = \frac{1 + m + \sqrt{1 + 2m}}{m} \qquad (10\text{-}23)$$

$$R_1 = \frac{R_o(a - 1)}{a + 1} \qquad (10\text{-}24)$$

$$R_2 = \frac{2aR_o}{a^2 - 1} \qquad (10\text{-}25)$$

Recall that m in Equations 10-22 and 10-23 is defined as $m = R_2/R_1$. One additional equation will be employed in this experiment so that the attenuator's characteristic resistance (R_o) can be determined from measured values. Specifically,

$$R_o = \sqrt{R_{in(o)}\, R_{in(s)}} \qquad (10\text{-}6)$$

Recall that the characteristic resistance of an attenuator is the value of R_L that makes the input resistance of the attenuator equal to R_L.

Procedure

Step 1. Calculate the characteristic resistance (R_o) and attenuation (a) for the attenuators in Figs. 10-26A and 10-26B.

$R_{o1} = \rule{2.5cm}{0.4pt}$ $\qquad\qquad$ $R_{o2} = \rule{2.5cm}{0.4pt}$

$a_1 = \rule{2.5cm}{0.4pt}$ $\qquad\qquad$ $a_2 = \rule{2.5cm}{0.4pt}$

Step 2. Measure the input resistance of each attenuator for the following values of R_L: 0, ∞, and 600 Ω. Record your data in Table 10-4.

Table 10-4. Measured Input Resistance

Attenuator 1		Attenuator 2	
R_L	R_{in}	R_L	R_{in}
0 Ω		0 Ω	
∞ Ω		∞ Ω	
600 Ω		600 Ω	

Step 3. Use Equation 10-6 to calculate R_o from the measured values in Step 2.

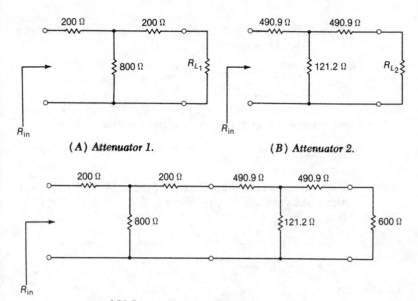

(A) Attenuator 1. (B) Attenuator 2.

(C) Input resistance of two 600-Ω sections.

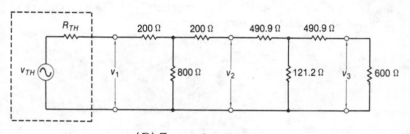

(D) Two-section attenuator.

Fig. 10-26. Schematic diagrams for Experiment 10-1.

$R_{o1} = $ _____ $R_{o2} = $ _____

Step 4. Cascade the two attenuator sections *as shown* in Fig. 10-26C. Measure the input resistance, R_{in}.

$R_{in} = $ _____

What would R_{in} equal if a number of 600-Ω sections were cascaded, assuming the last section was terminated in 600 Ω? Why?

Step 5. Obtain a signal generator whose output impedance is specified as 600 Ω. Adjust the output of the generator to 5 V peak (sinusoidal) at 1 kHz. Since you are measuring the generator's output voltage under *no-load* conditions, 5 V peak represents the Thevenin equivalent voltage.

289

Step 6. Connect the signal generator to the two-section attenuator in Fig. 10-26D. Measure v_1, v_2, and v_3.

$v_1 = $ _____ $v_2 = $ _____

$v_3 = $ _____

Step 7. From the *data* in Step 6 determine the attenuation of each section and the total attenuation from source to load.

$a_1 = $ _____ $a_2 = $ _____

$a_3 = $ _____

Step 8. Compare the measured values obtained in Steps 6 and 7 with calculated values for v_1, v_2, and v_3. You can calculate these voltages using the values of attenuation calculated in Step 1.

$v_1 = $ _____ $v_2 = $ _____

$v_3 = $ _____

Conclusion

How did the calculated and measured values compare? What is the relationship between total attenuation and the attenuation of each section in a multisection attenuator?

Selected British (American) Units and Conversion Factors

Table A-1. Common British Units

Quantity	Unit Name	Unit Symbol
length	foot	ft
mass	pound mass	lbm
time	second	s
force	pound force	lbf
charge	coulomb	C
temperature	degree rankine	°R
luminous intensity	candle	cd

Due to the fact that mass and force in the British (American) system, each employ the pound as a unit of measure some confusion may arise. *Mass* is a measure of the *amount* of material in a body. *Weight* is defined as the *force* with which a body is attracted towards the center of the earth. An object whose mass is one pound (1 lbm) weighs one pound (1 lbf) *only* at sea level. If you employ Newton's second law in the form $W = mg$, the force W (weight) has units of *poundals*, where

$$1 \text{ poundal} = \frac{1 \text{ lbm} \times \text{ft}}{s^2}$$

Thus, to distinguish between pound (mass) and pound (force) we employ the subscripts m and f, respectively (see Table A-1 and Chart

A-1). The term "weight" is often used "loosely" to mean mass since standards of mass (called weights) are often employed in the laboratory.

Chart A-1. Selected Conversion Factors

Length	12 in = 1 ft
	1 yd = 3 ft
	1 mi = 5 280 ft
	1 in = 2.54 cm
Mass	16 oz = 1 lbm = 0.453 6 kg
	1 ton = 2 000 lbm
Temperature	$1°F = 1°R$
	$°F = 1.8°C + 32$
Energy	1 Btu = 777.65 ft·lbf = 1 054.35 J
	1 kW·hr = 3.6×10^6 J
Power	1 hp = 1.98×10^6 ft·lbf = 745.7 W
Area	1 ft^2 = 929 cm^2
Force	1 lbf = 4.448 N

Scientific Notation

In science and technology you will encounter very large numbers and very small numbers. Scientific notation is extremely useful for representing these kinds of numbers.

In scientific notation a large or small number (N) is written as a number between 1 and 9.999 999 (M) multiplied by 10 to some power (n). Thus

$$N = M \times 10^n$$

where

N = number in standard notation,

$M \times 10^n$ = number in scientific notation.

The number n is the number of places the decimal point *of N* must be moved to the left or right to produce M. If $N < 1$, then the decimal point is moved to the right, making n negative $(0.002 = 2 \times 10^{-3})$. Similarly, if $N > 1$, then the decimal point is moved to the left, making n positive $(200 = 2 \times 10^2)$. If the decimal point does not have to be moved, then n is zero $(2 = 2 \times 10^0)$. The following examples illustrate the value of scientific notation.

Standard Notation	*Scientific Notation*
1	1×10^0
−1	-1×10^0
231	2.31×10^2
1 050	1.05×10^3
0.005	5×10^{-3}
0.000 000 000 000 150	1.5×10^{-13}
125 000 000 000 000	1.25×10^{14}

The last two examples above clearly demonstrate the utility of scientific notation. Multiplication and division of numbers written in scientific notation can be performed by following the rules below:

$$(A \times 10^x)(B \times 10^y) = AB \times 10^{x+y}$$

$$\frac{A \times 10^x}{B \times 10^y} = \frac{A}{B} \times 10^{x-y}$$

The examples which follow illustrate the use of these rules.

$$(10 \times 10^2)(5 \times 10^3) = 10(5) \times 10^{2+3}$$
$$= 50 \times 10^5$$
$$= 5 \times 10^6$$

$$\frac{2.5 \times 10^6}{1.25 \times 10^2} = \frac{2.5}{1.25} \times 10^{6-2}$$
$$= 2 \times 10^4$$

$$\frac{0.005(5\,000\,000)}{12.5} = \frac{(5 \times 10^{-3})(5 \times 10^6)}{12.5 \times 10^0}$$
$$= \frac{25 \times 10^3}{12.5 \times 10^0}$$
$$= 2 \times 10^3$$

$$\frac{120\,\text{mV}}{60\,\text{k}\Omega} = \frac{120 \times 10^{-3}\,\text{V}}{60 \times 10^3\,\Omega} = 2 \times 10^{-6}\,\text{A} = 2\,\mu\text{A}$$

$$(150\,\mu\text{A})(10\,\text{M}\,\Omega) = (150 \times 10^{-6}\,\text{A})(10 \times 10^6\,\Omega)$$
$$= 1\,500 \times 10^0\,\text{V}$$
$$= 1.5 \times 10^3\,\text{V}$$
$$= 1.5\,\text{kV}$$

The last two examples illustrate the use of a slight variation of scientific notation($M > 9.999\,999$) in conjunction with some of the prefixes introduced in Chapter 1. The reader is encouraged to become proficient with this notation.

Current Division
and Loop Equations

CURRENT DIVISION

Consider the circuit illustrated in Fig. C-1. The total current (I_T) divides between the two parallel branches consisting of R_1 and R_2. Thus

$$I_T = I_1 + I_2 \qquad\qquad \text{(C-1)}$$

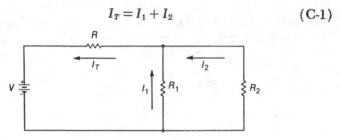

Fig. C-1. Current division.

Since R_1 and R_2 are connected in parallel we can write

$$V_{R1} = V_{R2}$$
$$I_1 R_1 = I_2 R_2$$
$$I_1 = \frac{I_2 R_2}{R_1}$$

Substituting the last expression into Equation C-1 yields

$$I_T = \frac{I_2 R_2}{R_1} + I_2$$

$$I_T = I_2\left(\frac{R_2}{R_1} + 1\right)$$

$$= I_2\left(\frac{R_1 + R_2}{R_1}\right)$$

Solving for I_2

$$I_2 = I_T\left(\frac{R_1}{R_1 + R_2}\right) \qquad \text{(C-2)}$$

Following a similar procedure for I_1

$$I_2 = \frac{I_1 R_1}{R_2}$$

$$I_T = I_1 + \frac{I_1 R_1}{R_2}$$

$$= I_1\left(\frac{R_1}{R_2} + 1\right)$$

$$= I_1\left(\frac{R_1 + R_2}{R_2}\right)$$

or

$$I_1 = I_T\left(\frac{R_2}{R_1 + R_2}\right) \qquad \text{(C-3)}$$

Equations C-2 and C-3 summarize the *current division* principle. This principle is very useful when working with shunted meter movements. Equations C-2 and C-3 should be memorized for future use!

LOOP EQUATIONS

Kirchhoff's voltage law states that the sum of the voltages around a closed path is zero. A *mesh* is defined as a loop that contains no loops within it. By assigning circulating currents (appropriately called mesh currents) to each mesh in a circuit we can write via Kirchhoff's voltage law an equation for each mesh. The solution of these mesh equations enables us to calculate the actual currents and voltages in the circuit. To illustrate the process we employ the following rules:

1. Assign counterclockwise or clockwise mesh currents to each mesh. Be *consistent,* that is, make *all* mesh currents either counterclockwise or clockwise—*don't* make some clockwise and some counterclockwise as this only serves to complicate the process needlessly.
2. Go around each mesh summing all voltages and equating the

resulting sum to zero. Remember that the voltage across a resistor equals the current through the resistor times the resistance of the resistor.

3. To help you implement Rule 2 when more than one mesh current flows through a resistor assume that *when in a given mesh that mesh current dominates*. This simply means that *when* you are writing the mesh equation for mesh 1 you will assume I_1 is larger than any other mesh current, *when* you are writing the mesh equation for mesh 2 you will assume I_2 is larger than any other mesh current, etc.

4. Solve the resulting mesh equations. Positive answers indicate the direction you initially assumed (counterclockwise or clockwise) was correct. Negative answers indicate the direction you initially assumed is incorrect and thus should be reversed.

5. Calculate from Rule 4 the actual currents and voltages.

To illustrate the process consider the circuit illustrated in Fig. C-2, where we have initially chosen to assign counterclockwise mesh currents.

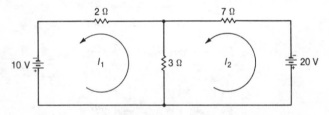

Fig. C-2. Mesh analysis.

Mesh 1—Starting at the positive terminal of the 10-V source, and proceeding around the mesh in the direction we have initially assumed I_1 to flow (counterclockwise) we have:

$$10 + 3(I_1 - I_2) + 2I_1 = 0$$
$$10 + 3I_1 - 3I_2 + 2I_1 = 0$$
$$5I_1 - 3I_2 = -10$$

Mesh 2—Starting at the top of the 3-Ω resistor we have:

$$3(I_2 - I_1) - 20 + 7I_2 = 0$$
$$3I_2 - 3I_1 - 20 + 7I_2 = 0$$
$$-3I_1 + 10I_2 = 20$$

Thus the mesh equations we wish to solve are

$$5I_1 - 3I_2 = -10$$
$$-3I_1 + 10I_2 = 20$$

297

After the necessary algebra we find

$$I_1 = -0.976 \text{ A}$$
$$I_2 = 1.71 \text{ A}$$

The negative answer for I_1 tells us our initial assumption for the direction of I_1 (counterclockwise) is incorrect. Thus we reverse the direction of I_1 as illustrated in Fig. C-3. With reference to Fig. C-3, it is apparent that the actual current through each resistor is as follows:

$$I_{2\Omega} = 0.976 \text{ A}$$
$$I_{3\Omega} = 0.976 \text{ A} + 1.71 \text{ A} = 2.686 \text{ A}$$

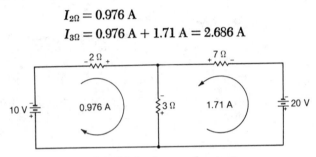

Fig. C-3. Solving for actual currents.

Similarly, the voltages across each resistor are:

$$V_{2\Omega} = (0.976 \text{ A})(2 \text{ }\Omega) = 1.952 \text{ V}$$
$$V_{7\Omega} = (1.71 \text{ A})(7 \text{ }\Omega) = 11.97 \text{ V}$$
$$V_{3\Omega} = (2.686 \text{ A})(3 \text{ }\Omega) = 8.058 \text{ V}$$

AC Electricity and RC Transients

Fig. D-1 illustrates the most common type of ac voltage—the sine wave. We will begin by introducing the notation, definitions, and conventions used to describe this type of waveform. With reference to Fig. D-1 we can define the following terms:

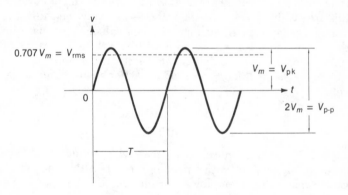

Fig. D-1. A sinusoidal voltage.

period (T)—This is the amount of time required to complete one cycle.

frequency (f)—This is the number of complete cycles occurring in one second. Frequency is related to the period as given by the following formula:

$$f = \frac{1}{T}$$

The unit of frequency is s^{-1}, or hertz (Hz).

peak value (V_m)—This is the maximum amplitude of the waveform. Since a sine wave is symmetrical about the horizontal axis the positive peak equals the negative peak.

peak-to-peak value ($V_{p\text{-}p}$)—As you can see in Fig. D-1 this is defined as follows: $V_{p\text{-}p} = 2V_m$.

average value (V_{avg})—This is the area under a waveform divided by the base. For a periodic waveform the base is taken to be the period. Since a sine wave is symmetrical about the horizontal axis the positive area equals the negative area resulting in a *true* average of zero. Thus $A_{avg} = 0$.

radian frequency (ω)—The radian frequency is defined as $\omega = 2\pi f$. The units of radian frequency are radians per second.

rms value (V_{rms})—The rms or *effective* value of a periodic voltage waveform is defined as follows:

$$V_{rms} = \sqrt{\frac{1}{T} \int_0^T v^2 \, dt} \qquad (D\text{-}1)$$

The symbol $\int$ tells us to *integrate* the voltage squared. Integration is a summing process that yields the area under the v^2 curve. Since this area is divided by the base, the term inside the radical represents a special type of average or *mean* value. The square root of this mean value *is* the rms value. What is the significance of the rms value? *The rms value is the equivalent amount of dc with respect to the heating effect produced.* For a *sinusoidal* current or voltage

$$I_{rms} = 0.707 I_m \qquad (D\text{-}2)$$
$$V_{rms} = 0.707 V_m \qquad (D\text{-}3)$$

Thus a 10-A-peak sinusoidal current will produce heat when it flows through a resistance equivalent to 7.07 A dc. The use of rms values enables us to employ for ac circuits the same power formulas that are used in dc circuits.

phase—The horizontal axis in Fig. D-1 is time. Recall that one cycle is equivalent to 360° or 2π radians. Thus you will encounter sketches of periodic waveforms where the units of the horizontal axis are either degrees or radians. Sinusoidal waveforms can differ from each other in either *amplitude, frequency,* or *phase,* or some combination of the three. For our purposes we will assume frequency is constant. Thus when comparing sinusoids we would expect to encounter differences in either amplitude and/or phase. Phase is simply the number of degrees or radians that one wave *leads* or *lags* another wave. Normally we express phase differences in degrees. A sinusoidal voltage is described by the following equation:

$$v = V_m \sin(\omega t \pm \theta) \qquad \text{(D-4)}$$

where

V_m = peak voltage,
$\omega = 2\pi f$ = radian frequency,
θ = phase angle.

Figure D-2 illustrates three sinusoids that differ only in phase. We will assume each sinusoid has a peak value of 10 V and a frequency of 1 kHz. The angle θ in Figs. D-2B and D-2C is assumed to be 80°. The equations that describe each of the sinusoids in Fig. D-2 are as follows:

$$\omega = 2\pi f = 2\pi(1\,000) = 6\,280 \text{ rad/s}$$

Thus

$$v_1 = 10 \sin 6\,280t$$
$$v_2 = 10 \sin(6\,280t - 80°)$$
$$v_3 = 10 \sin(6\,280t + 80°)$$

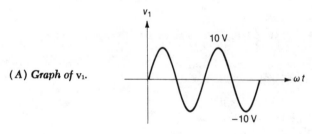

(A) Graph of v₁.

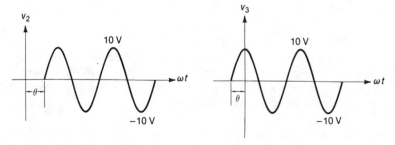

(B) Graph of v₂.   (C) Graph of v₃.

Fig. D-2. Phase differences.

phasors—A *phasor* is a rotating vector. Phasor diagrams are very useful when comparing sinusoids of the *same* frequency. In phasor
diagrams it is understood that the phasors are rotating counterclockwise at ω rad/s. Fig. D-3 illustrates a phasor diagram for the
sinusoids in Fig. D-2. Note in Fig. D-3 we have employed *rms
values* for each voltage, and uppercase letters to indicate the

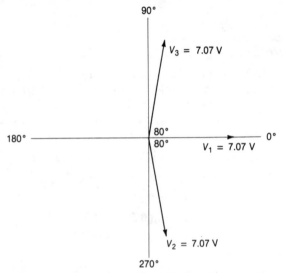

Fig. D-3. Phasor diagram for the sinusoids in Fig. D-2.

phasor quantities. From Fig. D-3 it should be apparent that we
can make the following statements:

V_3 leads V_1 by 80°
V_1 lags V_3 by 80°
V_1 leads V_2 by 80°
V_2 lags V_1 by 80°
V_3 leads V_2 by 160°
V_2 lags V_3 by 160°

Thus phase information is much easier to "see" from the phasor
diagram in Fig. D-3 than the sketches in Fig. D-2. In filter applications in control systems phase information is very important because phase differences are often used to turn load devices on or
off.

RC charging/discharging circuits—Capacitance has the property of
opposing changes in voltage. Fig. D-4A illustrates an *RC* charging circuit. Assuming the capacitor was initially uncharged, once
the switch is closed the voltage across the capacitor is given by

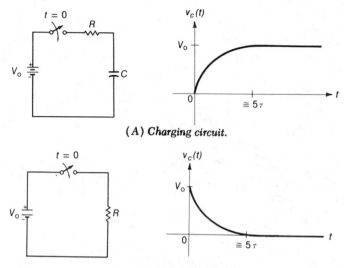

(A) Charging circuit.

(B) Discharging circuit.

Fig. D-4. RC charging and discharging.

$$v_C(t) = V(1 - e^{-t/\tau}) \qquad \text{(D-5)}$$

where

V = battery voltage in volts,
e = 2.718,
τ = time constant RC in units of seconds,
t = time in seconds.

The capacitor voltage charges *exponentially* to the battery voltage V. For our purposes we will assume the capacitor is fully charged after *five time constants* have elapsed. The point of all this is that capacitor voltage *cannot* change instantaneously. Thus in circuits containing capacitance a finite amount of time must pass after opening or closing switches for the capacitor voltage to reach its final value.

Fig. D-4B illustrates an *RC* discharging circuit. In this example the capacitor has been charged to V_o *prior* to closing the switch. Once the switch is closed the voltage across the capacitor is given by

$$v_C(t) = V_o e^{-t/\tau} \qquad \text{(D-6)}$$

As you can see in Fig. D-4B the capacitor voltage decays exponentially towards zero. Once again it takes approximately five time constants to reach the final value.

Thevenin's Theorem

Frequently when you encounter a circuit of moderate complexity you do not wish to determine the current through, voltage across, and power dissipated by each element in the circuit. You are primarily interested in determining the response associated with a particular element called the *load*. In simple terms Thevenin's theorem states:

As far as the load is concerned the rest of the circuit acts like an equivalent voltage source V_{TH} in series with an equivalent resistance R_{TH}.

This concept is illustrated in Fig. E-1. In Fig. E-1A the details of the original circuit are not shown. In effect the original circuit is

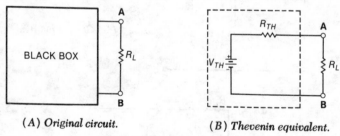

(A) *Original circuit.* (B) *Thevenin equivalent.*

Fig. E-1. Thevenin's theorem.

contained in an imaginary "black box." The load in Fig. E-1A is the resistance R_L connected between terminals A and B. Thevenin's theorem enables you to picture the contents of the black box as illustrated in Fig. E-1B. If the details of the original circuit are provided, how can you determine V_{TH} and R_{TH}? The following *rules* will permit you to achieve this modest objective.

Calculating V_{TH}:
1. Remove the load.
2. $V_{TH} = V_{oc}$, where V_{oc} is the "open-circuit voltage." This is simply the voltage between the terminals where the load has been removed.

Calculating R_{TH}:
1. Remove the load.
2. Reduce all sources to *zero*. This means you will replace ideal voltage sources by shorts, and ideal current sources by opens.

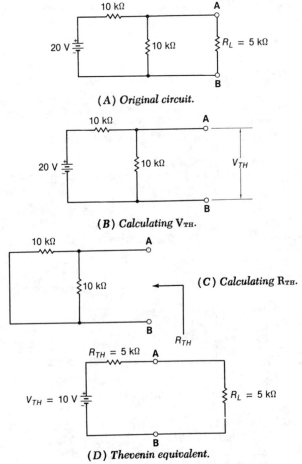

(A) *Original circuit.*

(B) *Calculating V*TH.

(C) *Calculating R*TH.

(D) *Thevenin equivalent.*

Fig. E-2. Circuits for Example C-1.

(If the sources are not ideal, they should be replaced by their internal resistances.)

3. R_{TH} is the resistance "looking back" into the circuit, from the load terminals, once Rules 1 and 2 have been implemented.

The following examples illustrate the application of the rules given above.

EXAMPLE E-1

Calculate R_{TH} and V_{TH} for the circuit shown in Fig. E-2. How much current flows through R_L? To calculate V_{TH} you remove the

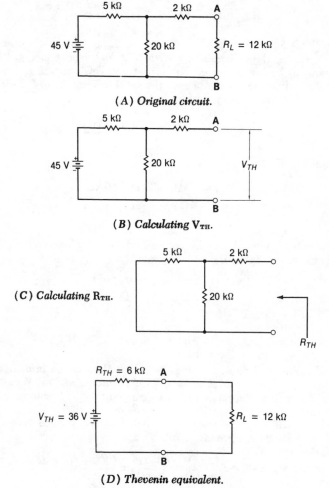

(A) *Original circuit.*

(B) *Calculating V*TH.

(C) *Calculating R*TH.

(D) *Thevenin equivalent.*

Fig. E-3. Circuits for Example C-2.

load and then determine the open-circuit voltage. Thus, from Fig.
E-2B,

$$V_{TH} = V_{oc} = \frac{(20 \text{ V}) (10 \text{ k}\Omega)}{20 \text{ k}\Omega} = 10 \text{ V}$$

To calculate R_{TH} you remove the load and reduce the 20-V source
to zero. Thus, from Fig. E-2C,

$$R_{TH} = 10 \text{ k}\Omega \parallel 10 \text{ k}\Omega = 5 \text{ k}\Omega$$

The current through R_L can be determined from the Thevenin
equivalent circuit in Fig. E-1D:

$$I_{RL} = \frac{V_{TH}}{R_{TH} + R_L} = \frac{10 \text{ V}}{5 \text{ k}\Omega + 5 \text{ k}\Omega} = \frac{10 \text{ V}}{10 \text{ k}\Omega} = 1 \text{ mA}$$

EXAMPLE E-2

Determine the voltage across R_L in Fig. E-3. Notice in Fig. E-3B
that with the load removed no current can flow through the 2-kΩ
resistor. Thus the voltage across the 2-kΩ resistor in Fig. E-3B is
zero, and $V_{oc} = V_{20k\Omega}$. Therefore

$$V_{TH} = V_{oc} = V_{20k\Omega} = \frac{(45 \text{ V}) (20 \text{ k}\Omega)}{25 \text{ k}\Omega} = 36 \text{ V}$$

From Fig. E-3C

$$\begin{aligned} R_{TH} &= 2 \text{ k}\Omega + 20 \text{ k}\Omega \parallel 5 \text{ k}\Omega \\ &= 2 \text{ k}\Omega + 4 \text{ k}\Omega \\ &= 6 \text{ k}\Omega \end{aligned}$$

The voltage across R_L can be determined from the Thevenin equivalent circuit in Fig. E-3D:

$$V_{RL} = \frac{(36 \text{ V}) (12 \text{ k}\Omega)}{6 \text{ k}\Omega + 12 \text{ k}\Omega} = \frac{(36 \text{ V}) (12 \text{ k}\Omega)}{18 \text{ k}\Omega} = 24 \text{ V}$$

EXAMPLE E-3

Assume the ammeter in Fig. E-4 has an $R_m = 100 \ \Omega$. Determine
the current through the ammeter.

The circuit illustrated in Fig. E-4A is a *Wheatstone bridge*. This
is an *important* circuit. Thus you should "internalize" the results of
this example problem so that you can quickly Theveninize other
Wheatstone bridge circuits. The open-circuit voltage V_{AB} in Fig.
E-4B equals V_{TH}. This voltage is equal to the voltage across R_2
minus the voltage across R_4. Thus

$$\begin{aligned} V_A &= V_{R2} = \frac{(300 \text{ V}) (400 \ \Omega)}{400 \ \Omega + 400 \ \Omega} = \frac{(300 \text{ V}) (400 \ \Omega)}{800 \ \Omega} \\ &= 150 \text{ V} \end{aligned}$$

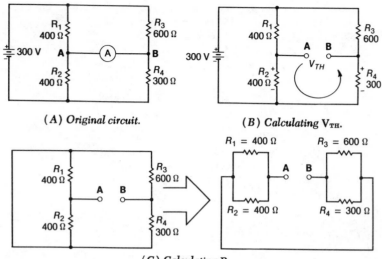

(A) Original circuit.

(B) Calculating V_{TH}.

(C) Calculating R_{TH}.

Fig. E-4. Solving for V_{TH} and R_{TH} in Example C-3.

$$V_B = V_{R4} = \frac{(300 \text{ V})(300)}{600 \text{ }\Omega + 300 \text{ }\Omega} = \frac{(300 \text{ V})(300 \text{ }\Omega)}{900 \text{ }\Omega}$$
$$= 100 \text{ V}$$

$$V_{AB} = V_{TH} = V_A - V_B$$
$$= 150 - 100 = 50 \text{ V}$$

The equivalent resistance between A and B in Fig. E-4C is R_{TH}. Notice that the circuit illustrated in Fig. E-4C has been redrawn to venin equivalent circuit illustrated in Fig. E-5. Thus

$$R_{TH} = R_{AB} = 400 \text{ }\Omega \parallel 400 \text{ }\Omega + 600 \text{ }\Omega \parallel 300 \text{ }\Omega$$
$$= 200 \text{ }\Omega + 200 \text{ }\Omega$$
$$= 400 \text{ }\Omega$$

The desired ammeter current is now easily determined from the Thevenin equivalent circuit illustrated n Fig. E-5. Thus

$$I_m = \frac{V_{TH}}{R_{TH} + R_m} = \frac{50 \text{ V}}{400 \text{ }\Omega + 100 \text{ }\Omega} = \frac{50 \text{ V}}{500 \text{ }\Omega}$$
$$= 0.1 \text{ A}$$

For any Wheatstone bridge circuit you can determine R_{TH} and V_{TH} as follows:

$$R_{TH} = R_1 \parallel R_2 + R_3 \parallel R_4 \tag{E-1}$$
$$V_{TH} = V_{oc} = V_{R2} - V_{R4}$$

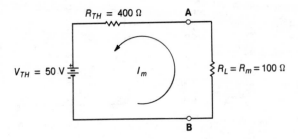

Fig. E-5. Thevenin equivalent circuit for Example C-3.

$$V_{TH} = V\left(\frac{R_2}{R_1 + R_2}\right) - V\left(\frac{R_4}{R_3 + R_4}\right) \qquad \text{(E-2)}$$

Equations E-1 and E-2 should be *memorized*. For quickly analyzing a circuit, Thevenin's theorem is one of the *most useful* tools you will ever encounter!

Answers to Odd-Numbered Problems

CHAPTER 1

1. 0.023 mV, 1 300 000 μV, 0.000 015 μF, 0.085 kΩ, 236 MΩ, 1.212 mA
3. 49.89 kg
5. 0.006 300 in
7. 1 kV, 100 V, 31.25 nA, 40 mA, 2 500 mV or 2.5 V, 100 MΩ
9. 5 820.1 tons

CHAPTER 2

1. (a) 0.02 mA
   (b) 0.88 mA to 0.92 mA, 0.38 mA to 0.42 mA
   (c) 2.22%, 5%
3. (a) A wood frame would *not* provide any electrical damping.
   (b) Since the spiral springs are used to supply current to the meter movement *both* are necessary.
   (c) Flush pole pieces would *not* concentrate the magnetic field in the space occupied by the moving coil as well as curved pole pieces. Thus physically larger magnets would be required.
   (d) Hard iron would become permanently magnetized the first time current flowed through the coil. Obviously this condition is undesirable.

CHAPTER 3

1. $R_m = 1$ kΩ
3. $a = 68.9\%$, $e = 31.1\%$

5. 34.42%

7. $a = 80\%, e = 20\%$

9. $R_1 = 2.22\ \Omega,\ R_2 = 20\ \Omega,\ R_3 = 88.8\ \Omega$

CHAPTER 4

1. $10\ \mathrm{k}\Omega,\ 50\ \mathrm{k}\Omega,\ 250\ \mathrm{k}\Omega,$ and $1\ \mathrm{M}\Omega$

3. Refer to Fig. 4-7. In this case $R_m = 100\ \Omega$, which can be neglected. Also note that a 0.1-V range is *not* specified as part of the design. Thus $R_1 = 750\ \mathrm{k}\Omega,$ $R_2 = 200\ \mathrm{k}\Omega,\ R_3 = 40\ \mathrm{k}\Omega,$ and $R_4 = 10\ \mathrm{k}\Omega.$

5. $8\ \mathrm{V},\ 0.47\ \mathrm{V}$

7. $a = 95.7\%, e = 4.3\%$

9. $V_{TH} = 40\ \mathrm{V},\ R_{TH} = 6\ \mathrm{k}\Omega,\ R_{1n}\ (\min) = 594\ \mathrm{k}\Omega$

CHAPTER 5

1. (a) $R_B = 29\ \mathrm{k}\Omega$
   (b) $\infty\ \Omega,\ 90\ \mathrm{k}\Omega,\ 30\ \mathrm{k}\Omega,\ 10\ \mathrm{k}\Omega,$ and $0\ \Omega$
   (c) $R_s = 64.34\ \mathrm{k}\Omega$

3. (a) $R_m' = 60\ \mathrm{k}\Omega,\ I_{FS} = 100\ \mu\mathrm{A}$
   (b) Fig. F-1

5. $363\ \Omega$

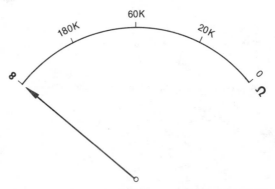

Fig. F-1. Calibrated ohms scale from Problem 5-3(b).

CHAPTER 6

1. Dc voltmeter: $R_{1n} = 10\ \mathrm{k}\Omega,\ 50\ \mathrm{k}\Omega,$ and $100\ \mathrm{k}\Omega$
   Hw voltmeter: $R_{1n} = 4.5\ \mathrm{k}\Omega,\ 22.5\ \mathrm{k}\Omega,$ and $45\ \mathrm{k}\Omega$
   Fw voltmeter: $R_{1n} = 9\ \mathrm{k}\Omega,\ 45\ \mathrm{k}\Omega,$ and $90\ \mathrm{k}\Omega$

3. $R_s = 44.5\ \mathrm{k}\Omega$

5. $R_1 = 45\ \mathrm{k}\Omega,\ R_2 = 33.75\ \mathrm{k}\Omega,\ R_3 = 10.25\ \mathrm{k}\Omega$

7. This is a good problem for "makeup" exams. The voltmeter will *not* read accurately. To understand why and provide additional "insight" the solution is cheerfully provided as follows. The *true-rms value* of the input voltage is $V_m/\sqrt{2} = 0.707V_m,$ or

$$V_{rms} = 0.707\,(14.14\text{ V}) = 10\text{ V}$$

When the voltage in Fig. 6-21E is measured by the voltmeter in Fig. 6-21C the following *average current* results:

$$I_{avg} = 0.636\left(\frac{V_m}{R_{in}}\right) = \frac{0.636(14.14\text{ V})}{45\text{ k}\Omega}$$

$$= 0.2\text{ mA}$$

The multiplying factor of 0.636 is used because the diode will always be forward biased by the input voltage (ideal diode). Recall that for a half-wave, rectifier-type, ac voltmeter the scale has been calibrated according to

$$V_{rms} = 2.22 R_{in}\,I_{avg}$$

Thus

$$V_{rms} = 2.22(45\text{ k}\Omega)(0.2\text{ mA}) = 20\text{ V}$$

This is the value *you would read*. Thus the voltmeter provides a measured value which is *twice* the true rms value. In order to flex your mental muscles rework Problem 7 assuming a full-wave, rectifier-type, ac voltmeter is employed for the measurement! This is an even better problem for a "makeup" exam!

CHAPTER 7

1. 10 μF
3. 50 μF
5. 395 kΩ, 20 V
7. $R_s = 248$ kΩ, $C = 0.04$ μF
9. $R_s = 98$ kΩ, $C = 0.1$ μF
   See Fig. 7-8. Scales 1 and 2 remain unchanged. Reading from left to right, scale 3 should read 0, 0.5, 1, 1.5, 2, 2.5, 3, 3.5, 4, 4.5, and 5 V.

CHAPTER 8

1. (a) 16 kΩ, (b) $V_{AB} = 1.006$ mV, $I_{AB} = 10.06$ μA
3. (a) 4 kΩ, (b) -114 mV, (c) $+111$ mV
5. 270°
7. $R_1 = R_3 = 499.999$ Ω, $R_2 = R_s = 8\,419.9$ Ω
9. Assuming $v_{o(max)} = \frac{2}{3}(9\text{ V}) = 6$ V: $R_s = 5.9$ kΩ and $I = 0.24$ mA

CHAPTER 9

1. $f_o = 33.86$ Hz, $|K_{PB}| = 1$
3. $\phi = -70.46°$
5. Approximately 16 V
7. Many solutions are possible. To illustrate the method:

$$R_{THC} = \frac{1}{2\pi \text{fC}} = 0.159 \times 10^{-3}\text{ s}$$

$$R_{TH} = \frac{(600\ \Omega + R)(1\ k\Omega)}{R + 1.6\ k\Omega}$$

Selecting $R_{TH} = 500\ \Omega$ and solving for R yields
$R = 400\ \Omega$. Thus

$$C = \frac{0.159 \times 10^{-3}}{5 \times 10^2} = 0.318\ \mu F$$

9. Again many solutions are possible. Example 9-7 illustrates the design process.

CHAPTER 10

1. $R_1 = 102.47\ \Omega$, $R_2 = 138.61\ \Omega$
3. $R_1 = 37.38\ \Omega$, $R_2 = 50.60\ \Omega$. See Fig. 10-22A for the appropriate conversion.
5. $48.45\ \Omega\ (R_{TH} = 0\ \Omega)$, $50\ \Omega\ (R_{TH} = 50\ \Omega)$, and $51.5\ \Omega\ (R_{TH} = \infty\ \Omega)$
7. $R_1 = 4\ M\Omega$, $C_1 = 11.75\ pF$
9. $R_1 = 5\ M\Omega$, $R_2 = 4.5\ M\Omega$, $R_3 = 0.5\ M\Omega$

Index